Cognitive Hyperconnected Digital Transformation Internet of Things Intelligence Evolution

RIVER PUBLISHERS SERIES IN COMMUNICATIONS

Indexing: All books published in this series are submitted to Thomson Reuters Book Citation Index (BkCI), CrossRef and to Google Scholar.

The "River Publishers Series in Communications" is a series of comprehensive academic and professional books which focus on communication and network systems. The series focuses on topics ranging from the theory and use of systems involving all terminals, computers, and information processors; wired and wireless networks; and network layouts, protocols, architectures, and implementations. Furthermore, developments toward new market demands in systems, products, and technologies such as personal communications services, multimedia systems, enterprise networks, and optical communications systems are also covered.

Books published in the series include research monographs, edited volumes, handbooks and textbooks. The books provide professionals, researchers, educators, and advanced students in the field with an invaluable insight into the latest research and developments.

Topics covered in the series include, but are by no means restricted to the following:

- Wireless Communications
- Networks
- Security
- Antennas & Propagation
- Microwaves
- Software Defined Radio

For a list of other books in this series, visit www.riverpublishers.com

Cognitive Hyperconnected Digital Transformation Internet of Things Intelligence Evolution

Editors

Ovidiu Vermesan

SINTEF, Norway

Joël Bacquet

EU, Belgium

Published, sold and distributed by:
River Publishers
Alsbjergvej 10
9260 Gistrup
Denmark

River Publishers
Lange Geer 44
2611 PW Delft
The Netherlands

Tel.: +45369953197
www.riverpublishers.com

ISBN: 978-87-93609-11-2 (Hardback)
       978-87-93609-10-5 (Ebook)

©2017 River Publishers

Dedication

"Imagination is not only the uniquely human capacity to envision that which is not, and therefore the fount of all invention and innovation. In its arguably most transformative and revelatory capacity, it is the power that enables us to empathize with humans whose experiences we have never shared."

– J. K. Rowling

"It's always about timing. If it's too soon, no one understands. If it's too late, everyone's forgotten."

– Anna Wintour

Acknowledgement

The editors would like to thank the European Commission for their support in the planning and preparation of this book. The recommendations and opinions expressed in the book are those of the editors and contributors, and do not necessarily represent those of the European Commission.

Ovidiu Vermesan
Joël Bacquet

Contents

Preface

The Internet of Things: Driving the Digital Transformation of the Global Economy

The Internet of Things (IoT) is a multidimensional and multi-faceted paradigm. It describes how many fixed and mobile "things" with various levels of "intelligence" are interconnected, sensing/actuating, interpreting, communicating and processing information and exchanging knowledge over the hyperconnected IoT space by using different types of platforms.

IoT will force business transformation and bring fundamental changes to individuals' and society's perspectives on how technologies and applications work in the world.

A citizen-centric IoT environment requires tackling new technological trends and challenges. This will have an important impact on research activities, which need to be accelerated without compromising the thoroughness and rigour of the testing and the time required for commercialisation.

A knowledge-centric network for IoT, IoT context-awareness, IoT traffic characterisation/monitoring/optimisation and the modelling and simulation of large-scale IoT scenarios have to be addressed in order to enable real-life large-scale deployments, testbeds, prototypes and practical systems for IoT use cases across industrial sectors.

The IoT is at the core of the digitisation processes of the economy and society and an essential building block for the Digital Single Market (DSM), which will have a potentially significant impact on Europe's prosperity and competitiveness in a globalised economy.

For Europe, a new dynamic and connected engine for research and innovation is needed in the area of IoT in order to maintain the EU's innovative spirit and global edge in IoT research and generate new jobs and sustainable economic growth.

The following chapters provide an overview of the state of the art for research and innovation in the era of convergence of industrial/business/consumer IoT, while presenting the challenges and opportunities for future IoT ecosystems and addressing IoT technologies, applications development and deployment for various domains and across domains.

Editors Biography

Dr. Ovidiu Vermesan holds a Ph.D. degree in microelectronics and a Master of International Business (MIB) degree. He is Chief Scientist at SINTEF Digital, Oslo, Norway. His research interests are in the area of mixed-signal embedded electronics and cognitive communication systems. Dr. Vermesan received SINTEF's 2003 award for research excellence for his work on the implementation of a biometric sensor system. He is currently working on projects addressing nanoelectronics, integrated sensor/actuator systems, communication, cyber-physical systems and the IoT, with applications in green mobility, energy, autonomous systems and smart cities. He has authored or co-authored over 85 technical articles and conference papers. He is actively involved in the activities of the Electronic Components and Systems for European Leadership (ECSEL) Joint Technology Initiative (JTI). He has coordinated and managed various national, EU and other international projects related to integrated electronics. Dr. Vermesan actively participates in national, H2020 EU and other international initiatives by coordinating and managing various projects. He is the coordinator of the IoT European Research Cluster (IERC) and a member of the board of the Alliance for Internet of Things Innovation (AIOTI).

Joël Bacquet is a senior official of DG CONNECT of the European Commission, taking care of the research and innovation policy for the Internet of Things. Before working in this field, he was programme officer in "Future Internet Experimental Platforms", head of the sector "Virtual Physiological Human" in the ICT for health domain. From 1999 to 2003, he was head of the sector "networked organisations" in the eBusiness unit. He started working with the European Commission in 1993, in the Software Engineering Unit of the ESPRIT Programme. He started his carrier as visiting scientist for Quantel a LASER company in San José, California in 1981. From 1983 to 1987, he was with Thomson CSF (Thales) as software development engineer for a Radar System. From 1987 to 1991, he worked with the

European Space Agency as software engineer on the European Space shuttle and international Space platform programmes (ISS). From 1991 to 1993 he was with Eurocontrol where he was Quality manager of an Air Traffic Control system. He is an engineer in computer science from Institut Supérieur d'Electronique du Nord (ISEN) and he has a MBA from Webster University, Missouri.

List of Figures

List of Tables

1

IoT Driving Digital Transformation – Impact on Economy and Society

Mechthild Rohen

European Commission, Belgium

1.1 IoT as a Major Enabler for Digitizing Industry

In recent years, the Internet of Things (IoT) has been gathering pace and unleashing a very disruptive potential. According to Gartner, nearly five billion "things" were connected in 2015 and the number will reach 50 billion by 2020. However, the IoT does not only have a disruptive power but it is also one of the main drivers and enablers for the Digitising European Industry strategy announced by the European Commission in April 2016.

IoT is as multidimensional and multi-faceted as the many 'things' that form it, therefore the main issues and challenges have to be addressed comprehensively and from many angles. At the European Commission, we are addressing IoT as a strategic dimension of the Digital Single Market (DSM), not only in terms of regulatory challenges but also with regard to interoperability issues and the possible fragmentation of standards, probably the most important obstacles at the moment. The IoT is therefore at the core of digitisation processes of the economy and society and an essential building block for the DSM.

A fully-functioning Digital Single Market is a pre-condition for Europe's prosperity and competitiveness in a globalised world economy. The DSM aims to remove boundaries and obstacles in the digital world and to boost our internal market of nearly 510 million customers and over 20 million companies for digital products.

1.2 Main Elements of the IoT Implementation Plan and Its First Pillar

The European Commission, being fully aware of the importance of IoT and its growing impact on EU citizens' lives and on the European economy, has been consistently aiming to achieve Leadership in Internet of Things for Europe, as envisaged in the Communication on 'Digitising the European Industry (DEI) – Reaping the full benefits of a Digital Single Market'.

All our actions aim at achieving 3 objectives, outlined in the Staff Working Document 'Advancing the Internet of Things in Europe': (i) a single market for IoT – which means seamlessly connected devices and services; (ii) a thriving IoT ecosystem – including open platforms and standards used across sectors; and – (iii) last but not least – a human-centred IoT – encompassing European values, personal data and security. These 3 objectives are at the same time the basis for the main pillars of our IoT implementation strategy.

Under the first pillar, our efforts aim to provide appropriate regulatory framework conditions to facilitate the creation of an IoT single market. In this context, the Commission is focusing on 5 specific actions. Given the huge importance of data ownership, privacy and security to EU citizens, we are clarifying the legal framework in relation to data, its free flow and ownership. This means that we are trying to answer the question – Who has the economic right to data? We are also dealing with rights to access, transfer and usage of non-personal data.

Additionally, the Commission is evaluating the current product liability framework for its appropriateness regarding the data-based IoT products and services, and autonomous systems. Our aim is to increase investment security for companies and to reassure consumers about their right to receive compensation in case of damages caused by using these emerging technologies. The need for adapting the current liability framework is currently under review following an open public consultation and an assessment of existing law.

On top of that, we are promoting an interoperable IoT numbering space for universal object identification that transcends geographical limits, as well as an open system for object identification and authentication. In September 2016, in the proposal for the recast of the European Electronic Communications Code, we have also proposed the harmonisation and clarification of rules and governance of numbering in the M2M context, as a response to the increasing demand for such numbering resources, in particular for IoT.

Furthermore, given the fact that 5G technology is an enabler of IoT, but not yet deployed, we aim at ensuring the availability of sufficient spectrum for alternative IoT communications.

Having recognised the need to increase the trust in IoT, we are revising our 2013 Cybersecurity strategy to take into account the IoT and we are exploring the possibility of a European ICT security certification framework including the IoT aspect. In autumn this year we will come back with the means to address certification but also potential labelling as integrated measure in an ICT security certification framework. A possible "Trusted IoT label" would provide the users with more transparency and inform them properly about the different levels of trust, privacy and security an IoT solution is offering. It would therefore allow them to make more credible and reliable choices.

1.3 The Second and the Third Pillar – Projects, Partnerships and Standardisation

The actions that are being undertaken under the second pillar of the IoT implementation plan focus on creating a thriving IoT ecosystem by supporting the IoT Large Scale Pilots (LSPs) and the Alliance for Internet of Things Innovation (AIOTI). LSPs are big projects that aim at fostering the deployment of IoT solutions in Europe through integration of advanced IoT technologies across the value chain. The LSPs will define an open architecture, platforms, interoperability and will be quite active in standardisation activities. We have launched 5 LSPs at the beginning of 2017 with H2020 financial support of €100 million in 5 areas: Smart living environments for ageing well; Smart Farming and Food Security; Wearables for smart ecosystems; Reference zones in EU cities; and Autonomous vehicles in a connected environment. We plan also to look into other sectors for the use of IoT, e.g. energy. This level of funding for a pilot implementation is of a large enough scale to achieve a critical mass.

Through the current ongoing LSPs, we are supporting elderly people by giving them access to healthcare from home. We are increasing agricultural productivity and food safety, and we are also decreasing energy consumption and pollution in cities. We are increasing security in public places, mobility and traffic efficiency through the connected cars. The CORDIS website [1] provides more information about the projects and the benefits they are bringing to the European citizens.

Moreover, in order to support the IoT deployment in Europe, we also launched in January 2016 nine projects supported by €53 million of EU H2020 funding. The aim of this European Platform Initiative (EPI) is to create ecosystems of "Platforms and Tools for Connected Smart Objects" and to overcome the fragmentation of vertically-oriented closed systems, architectures and application areas. Up to €10 million is envisaged for SMEs and start-ups working with these platforms.

One of the most prominent IoT stakeholder groups, which support technical and policy measures, is the Alliance for Internet of Things Innovation (AIOTI) that was launched by Commissioner Oettinger in March 2015 to assist the creation of a dynamic European IoT ecosystem.

The overall goal of the Alliance is the creation of a dense European IoT ecosystem to unleash the potential of the IoT. AIOTI is being consulted by the Commission on the hot topics in IoT. These include regulatory policy issues in order to help building consensus at a European level and to provide recommendations based on a broad stakeholder community. The Alliance has acquired a distinct legal personality by becoming an independent association under Belgian law in September 2016 and has almost 180 members from an industrial background, covering important areas, such as: aging well, smart farming, smart industries, smart cities, transportation and traffic, and wearables.

Of high importance is also the third pillar of our IoT implementation strategy – that is mainly focused on standardisation and interoperability.

The European Commission is fostering an interoperable environment for the IoT by closely cooperating with European and International Standards Organisations. We are continuously monitoring and reviewing the progress in promoting IoT standardisation and interoperability that would create a true IoT Digital Single Market. Additionally, we are promoting the uptake of IoT standards in public procurement.

Thanks to the tremendous work of the AIOTI, Europe has delivered on important milestones which were presented earlier this year in a joint workshop with the Commission. We gained an authoritative landscape of standardisation organisations and industrial alliances, covering all sectors and all technologies, which is being used and copied worldwide in developing standardisation strategies. Moreover, AIOTI has established itself at the nexus of global standardisation coordination and is establishing operational links with world leaders on IoT standardisation (IEEE, W3C, OneM2M, ISO/IEC, ITU, IETF, just to name a few).

The two major deliverables in this respect are the high level reference architecture, which is being used to develop interoperable IoT solutions and the joint White Paper produced by AIOTI, W3C, IEEE and oneM2M on Semantic Interoperability for the Web of Things published in December 2016.

Moreover, we are supporting international collaboration on IoT with strategic partner countries such as: Japan, South Korea, China, Brazil and USA. The cooperation takes place through policy dialogues and/or joint calls under the Horizon 2020.

Also, AIOTI has signed international agreements for industrial collaboration with similar entities in Japan and Brazil. At the Mobile World Congress 2017 in Barcelona AIOTI signed a Memorandum of Understanding with the Brazilian Camara do IoT and at CeBIT 2017 in Hannover a similar agreement with the Japanese IoT Acceleration Consortium.

1.4 Conclusion

In order to fulfil the ambitious goals outlined above, all stakeholders across the value chain have to be mobilised. We need to address new ways of value creation and enable the emergence of new cross-cutting business models. Promotion of open IoT ecosystems with special attention to SMEs and start-ups, as well as the creation of a market on IoT and a harmonised regulatory environment, especially in terms of object identification, connectivity and numbering, are of utmost importance.

The key instruments that can help in the implementation of the above steps are the Work Programme 2018–2020 for ICT in Horizon 2020 and the AIOTI.

To actively shape the IoT policy, companies, associations, Members States and other IoT stakeholders should combine and channel their resources. The European Commission can play a role as coordinator and facilitator in this process, helping to break silos and aiming to support European and global cooperation on IoT.

Reference

[1] CORDIS online at http://cordis.europa.eu/search/result_en?q=contenttyp e=%27project%27%20AND%20/project/relations/associations/related Call/call/identifier=%27H2020-IOT-2016%27

2

Next Generation IoT Platforms

Joël Bacquet and Rolf Riemenschneider

European Commission, Belgium*

Abstract

Industry is one of the pillars of the European economy – the manufacturing sector in the European Union accounts for 2 million enterprises, 33 million jobs and 60% of productivity growth. The new industrial revolution is driven by new-generation information technologies such as the Internet of Things (IoT), cloud computing, big data and data analytics, robotics and 3D printing. These technologies open new horizons for industry to become more adventurous, more efficient, to improve processes and to develop innovative products and services. Recent studies estimate that digitisation of products and services can add more than €110 billions of annual revenues in Europe in the next five years. In this context, Digitizing European Industry (DEI) initiative aims to establish, together with all Member States, and industry and other stakeholders, and building upon existing multi-stakeholder dialogues, a framework for facilitating coordination and cooperation of European, national and regional initiatives on digitising European industry, as well as to mobilise stakeholders across the value chains.

2.1 Introduction

The Communication on DEI was adopted last April 2016. The overall objective of the DEI initiative is to ensure that any industry in Europe, big or small, wherever situated and in any sector, can fully benefit from digital

---

*The views expressed in this article are purely those of the authors and not, in any circumstances, be interpreted as stating an official position of the European Commission.

innovations to upgrade its products, improve its processes and adapt its business models to the digital change. This required the full integration of digital innovations across all sectors of the economy. In the DEI strategy, the IoT objectives are indicated under the headings "Achieving Partnerships for leadership in digital technologies, value chains and platforms". In this chapter, the results of the related Working Group on "platforms" and more particularly the sub-group on "IoT platforms" will be presented. The activities started with the 1st roundtable in September and ended in May 2017 with a set of recommendations.

The ongoing digitisation of industry and services has a profound effect across all sectors. It is underpinned by research and innovation in relation to several technological trends. The Internet of Things, data analytics, cloud, high-performance computing and artificial intelligence are the most prominent ones. Advances in these technologies are transforming products, processes and business models and ultimately reshuffling global value chains in all sectors (Figure 2.1).

To maintain their competitive edge, companies have to fully embrace digitisation not only by making the best use of the latest digital technologies but also by integrating digital innovations as key elements of their development strategies. Next digital champions can emerge in any sector of the economy from construction and textile to health or energy equipment.

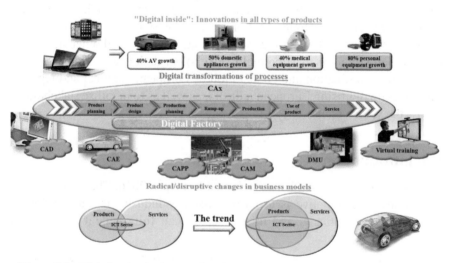

Figure 2.1   Digital technologies transforming products, processes and business models.

In Europe, many companies, especially in the high-tech sector, are already grasping the opportunities of this new industrial revolution, and studies show that digitisation of products and services could increase EU industry revenues by €110 billion a year.

However, many traditional sectors and small and medium-sized enterprises (SMEs) are lagging behind: less than 2% of them use advanced digital technologies to innovate in products and processes. In addition, a large disparity exists between Member States and regions creating a new 'digital divide', which can ultimately hurt all economies in Europe.

To tackle these challenges, the European Commission launched the Digitising European Industry [1] initiative in April 2016. Its overall objective is to put in place the necessary mechanisms to ensure that every industry in Europe, in whichever sector, wherever situated, and no matter of which size can fully benefit from digital innovations.

The initiative focuses on actions with a clear European value-added. It builds on the momentum and it complements and connects different national initiatives. Of particular importance are four action lines: the European platform of national initiatives including Digital Innovation Hubs, Digital Industrial Platforms, Digital Skills and Smart Regulation.

2.2 DEI Implementation – Working Groups

The implementation of the DEI initiative is supported by a Roundtable of high-level representatives of Member States' initiatives, industry leaders and social partners. To support the work of the Roundtable, two Working Groups (WG) have been set-up in order to make progress on aspects of the implementation of the DEI Action Plan:

- WG1: Mainstreaming digital innovations across all sectors;
- WG2: Strengthening leadership in digital technologies and in digital industrial platforms across value chains in all sectors of the economy.

The WGs performed fact finding, collected best practices and formulated recommendations, addressed to the High-Level Representatives attending the Roundtables. On industrial platforms, the challenge is to seize the opportunities arising from digitisation to establish European leadership in the next generation digital platforms and re-build the necessary, underlying digital supply chain on which all economic sectors are increasingly dependent. We addressed this challenge in the WG2 by pursuing the objectives to provide a view on the current landscape of platform activities, to develop a vison for the future European platforms, to analyse the gaps, to identify the key players and key initiatives and develop an action plan with recommendations to the roundtable. The WGs produced several reports available on Futurium [2].

To allow in depth analysis the Working Group 2 was further broken down into specific vertical areas (Smart Agriculture, Connected Smart Factory and Digital Transformation of Health and Care) and on horizontal issues (Industrial Data Platforms and Internet of Things) as depicted in Figure 2.2.

From now on we will focus on the results of the sub-working group on "IoT platforms".

2.3 IoT Platforms – State of Play

The term 'platform' has several different meanings. The DEI Communication defines it, rather narrowly, as "multi-sided market gateways creating value by enabling interactions between several groups of economic actors" [3]. A broader, and more useful, interpretation would be "agreements on functions and interfaces between industry players that create markets and market opportunities leading to ecosystems and standards." This definition could be instantiated for IoT with the following meaning. IoT platforms supported by dynamic ecosystems should cover the complete value chain from sensors and actuators, connectivity, cloud infrastructure, applications and services.

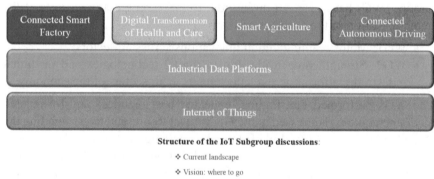

Figure 2.2   Digitising European Industry – working group 2 subgroups.

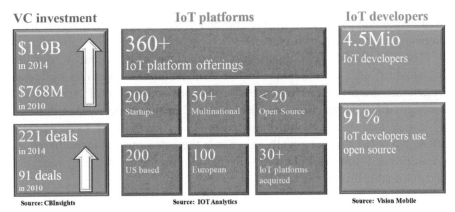

Figure 2.3   IoT platform landscape.

The current landscape of IoT <u>platforms</u> highlights that the European offer is very fragmented. The coordination and support action, UNIFY-IoT has identified that around 360 known platforms exist, with the vast majority of these being developed from 2013 onwards (Figure 2.3).

Across the 360 platforms identified world-wide there is a mix of cloud companies, some open source platforms, some industrial sector driven platforms, e.g. by Bosch, Siemens or GE, and some standards based solutions.

This burst of new platforms is mainly dominated by US suppliers. The initial explosion of new platforms supported by SMEs in the area is, however, slowing down which is a sign of a new immature market. There will thus be a natural selection within the market over the next few years where competition is fierce.

Internet giants like Google, Amazon or Apple have invested in traditional markets like transport (the Google car), smart home (the NEST thermostat) or big data for energy (Google Energyworx) – and this could cause a seismic shift in traditional sectors and it value chain from product design, development and service provision.

The risk of US dominance in future industrial platforms is also quite high and the above-mentioned large companies have significant resources to acquire a dominant position and impose de-facto standards.

It may be worth making a distinction between IoT for consumers and Industrial IoT with the first covering smart phones, fitness tracking tools etc. and the second being exploited in areas such as smart factories, smart health care, etc. However, the borderline between consumer and industry is blurred as machine data allow to extract personal information. Fast advances are being made in the consumer world and a question is how this can drive innovation in the industrial space.

The industrial IoT domain summarizes everything what is outside the classical consumer domain with a strong emphasis on B2B business. In general, there is a convergence of consumer and industrial Internet. We see signs of 'consumerization', for instance, in the home market through the appearance of voice control appliances like Amazon's Alexa or Apple's Homepad. Also, this is typically the case in the automotive industry in which consumer and industrial platforms are merging in the concept of connected and automated driving.

2.4 Needs and Priorities for the Next Generation IoT Platforms

As highlighted in 2.2 the current fragmentation of IoT platforms creates challenges for business and consumers. With the proliferation of platform offerings there is considerable fragmentation and at present there is no clear convergence towards a limited number of platforms. Although we may expect some form of consolidation in the future, there is a clear need for interoperability and a need for open platforms to foster up-scaling and to avoid vendor lock-in. This issue may require collaboration on the definition of common interfaces and data format. The emphasis should be on developing a convergence on existing standards rather than in generating new platforms and new IoT standards. A starting point of this collaboration could happen at different levels with different communities:

- IoT European Platforms Initiative (IoT-EPI) as part of the Cluster of Research and Innovation projects in which about 34 different platforms are demonstrated and used in different use cases;
- Use cases supported by the EU funded IoT European Large-Scale Pilots Programme. These actions should be complemented with the support to standardisation activities and the help of the AIOTI which has a major role to play in forging consensus in industry. A European approach to standards would eventually influence standards at an international level such as oneM2M or W3C in the case of defining semantic interoperability.

Besides a diversity of standards, European industry is confronted with increasing security risks. Cybersecurity continues to be a serious issue in the IoT space, especially with recent high-profile breaches, essentially through botnet attacks. Security and privacy need to be addressed at different levels. Privacy needs to be protected within each sector. Privacy and end-to-end security solutions should be addressed based on local reasoning and trust, validating novel business models when data is aggregated and shared across the value chains.

End-to-end security and security by design needs to be provided at all levels, considering devices, platforms and the connecting network. Beyond this, risk assessment and security up-dates should be continuously managed over time during the entire product lifecycle within a framework that involves all actors of the value chain. The next generation IoT platforms will have to be much more secured and support agile concepts to respond to dynamic security risks, whilst also ensuring the privacy of the users. Hopefully a starting point will be the results of the IoT-3 Call with a set of resulting projects that shall integrate sophisticated security concepts with emerging IoT architectures for ensuring end-to-end security in highly distributed, heterogeneous and dynamic IoT environments.

As more critical applications are connected via networks, e.g. autonomous driving or energy grids, there will be a need for higher quality network connections. If availability and timeliness in networks cannot be guaranteed there will be a need to keep processing local. Indeed, there is a growing trend towards data processing close to the point of action, i.e. edge computing, to address real-time availability in platforms and also limit liability and risk that would otherwise be incurred from performing processing in central clouds. New high potential areas using the recent advances in Artificial Intelligence could be explored to filter, analyse and process the data at the sensors/actuators level, with increasing computational power close to the network edge.

Much of the potential value from IoT will come from moving beyond proprietary technologies and sectoral siloes that largely exist today. New revenues may come from product and service innovations that enable economic growth and up-scaling of existing platforms across different sectors. The rise of the IoT will change the 'vertically integrated' economic models, and open up new horizons for industry to become more adventurous, more efficient, and to develop innovative products and services.

A firm strategy on digital transformation involving a strategy for leadership on IoT platforms needs to be supported by national programmes and policies. There are a number of national activities on platforms and it would be beneficial to share information between these. However, there are some challenges. Useful information on the impact of platforms is not yet available and platform development is being driven by different sectors, still in isolation. For instance, in France platform development is being driven by the micro-electronics industry and in the UK it is driven by smart city or manufacturing applications. This leads to differences in national IoT strategies. Actions that foster more explicit activities between member states are needed that spread best practice and also increase the awareness of IoT. A horizontal approach is needed to support convergence of IoT platforms and the benefits can be maximised through coordination at a European level via connection of regional and national innovation hubs.

To promote acceptance and prove the reliability of platforms there is also a need for large experimental facilities for testing and demonstration of novel standards, architectures and platforms driven by selected verticals. Here there is a key need to avoid silos and vendor lock-in. Open standards and open APIs are important elements to allow SMEs to access and exploit an IoT platform. A vibrant developer community is an essential element of an ecosystem supporting a platform. To ensure this it is important that small SMEs and start-up players get involved in these activities to address new business opportunities and business models.

2.5 Conclusion

Future platforms should bridge the current interoperability gap between the vertically-oriented IoT platforms and mobilise third party contributions by creating marketplaces for IoT services and applications. A harmonised European market for IoT interoperability standards and open APIs are a prerequisite for a free marketplace which reduces dependencies and barrier for new business and SMEs.

Security and privacy have to be addressed in priority in the next generation IoT platforms. This will require actions not only in research and development, in large scale deployment but also at policy level with the possibility to introduce IoT trusted labels.

Although there are significant investments in national IoT strategies, initiatives addressing platform building will not be able to compete at an international level as there is insufficient local user base. In order to compete there is a pressing need for Europe to co-ordinate activities to create critical mass and avoid fragmentation and silos.

Last but not least, there is a need to promote a dialogue across a critical mass of stakeholders, including large companies across the value chain but also across vertical sectors, and promote consensus on platform up-scaling.

As mentioned above, an open platform approach would be sought of, allowing start-ups and entrepreneurs to create new services, giving SMEs access to new technologies and emerging standards. This stakeholder dialogue can be supported by the alliance AIOTI with the aim of bringing different communities together. Here there is a need to discuss legal issues, technical bottlenecks and market barriers. The aim would be to drive the convergence of standards across different sectors and accelerate adoption of IoT platforms in relevant sectors (sector-specific) whilst promoting spill-over effects to other sectors.

The European Commission encourage industry to drive standards and platform building that enable cross-sector fertilisation and up-scaling. For this to happen, the Commission is prepared to support leadership in platforms through its WP2018-20 under Horizon 2020 by large testbeds allowing companies to collaborate on connected objects and test new business models across different sectors.

References

[1] https://ec.europa.eu/digital-single-market/en/digitising-european-industry?utm_source=twitter&utm_medium=social&utm_campaign=DigitisingIndustry
[2] Futurium: https://ec.europa.eu/futurium/en/dei-implementation/library
[3] COM (2016) 180 final, 19 April 2016
[4] IoT European Large-Scale Pilots Programme, online at https://european-iot-pilots.eu/

3

Internet of Things Cognitive Transformation Technology Research Trends and Applications

Ovidiu Vermesan[1], Markus Eisenhauer[2], Harald Sundmaeker[3], Patrick Guillemin[4], Martin Serrano[5], Elias Z. Tragos[5], Javier Valiño[6], Arthur van derWees[7], Alex Gluhak[8] and Roy Bahr[1]

[1]SINTEF, Norway
[2]Fraunhofer FIT, Germany
[3]ATB Institute for Applied Systems Technology Bremen, Germany
[4]ETSI, France
[5]Insight Centre for Data Analytics, NUI Galway, Ireland
[6]Atos, Spain
[7]Arthur's Legal B.V., The Netherlands
[8]Digital Catapult, UK

Abstract

The Internet of Things (IoT) is changing how industrial and consumer markets are developing. Robotic devices, drones and autonomous vehicles, blockchains, augmented and virtual reality, digital assistants and machine learning (artificial intelligence or AI) are the technologies that will provide the next phase of development of IoT applications. The combination of these disciplines makes possible the development of autonomous systems combining robotics and machine learning for designing similar systems. This new hyperconnected world offers many benefits to businesses and consumers, and the processed data produced by hyperconnectivity allows stakeholders in the IoT value network ecosystems to make smarter decisions and provide better customer experience.

3.1 Internet of Things Evolving Vision

IoT technologies and applications are creating fundamental changes in individuals' and society's view of how technologies and businesses work in the world. The IoT has changed the way that connected vehicles work, facilitating the functionalities with automated procedures. The IoT connects vehicle to vehicle, assisting with collision avoidance and vehicle to infrastructure, preventing unscheduled lane departure and automating toll collection. Vehicles are manufactured in a way that facilitates the employment of IoT technologies, with autonomous driving technology and features integrated into the vehicle, e.g. automatic and responsive cruise control based on recognizing and responding to traffic signs and communication with the infrastructure of the city (i.e., traffic lights, buildings, etc.). Maintaining digitally-connected lifestyles is supported by IoT technologies that are improving the physical driving experience, and make it more enjoyable by integrating it with new "mobility as a service" concepts and business models.

Citizen-centric IoT open environments require new technological trends and challenges to be tackled. In this context, future developments are likely to require new businesses, business models and investment opportunities, new IoT architectures and new concepts and tools to be integrated into the design and development of open IoT platforms. This becomes evident in scenarios where IoT infrastructures and services intersect with intelligent buildings that automatically optimize their HVAC and lighting systems for occupancy and reduced energy usage. Other examples include heavy machinery that predicts internal part failure and schedules its own maintenance or robotic and autonomous system technologies that deliver advanced functionality.

IoT is the result of heterogeneous technologies used to sense, collect, act, process, infer, transmit, notify, manage and store data. IoT includes also the combination of advanced sensing/actuating, communication, and local and distributed processing, which takes the original vision of the IoT to a wholly different level, opening up completely new classes of opportunities for IoT with many research challenges to be addressed spanning several research areas.

3.1.1 IoT Common Definition

IoT is transforming the everyday physical objects in the surrounding environment into ecosystems of information that enrich people's lives. IoT is bridging the gap between the physical and the digital or virtual worlds, facilitating the convergence of advances in miniaturization, wireless connectivity, increased data-storage capacity and batteries. IoT is a set of key enabling technologies for digital businesses and one of the main drivers contributing to transforming the Internet and improving decision-making capacity via its augmented intelligence. People will engage with IoT applications using all their senses: touch and feel, sight, sound, smell and taste, individually or in combination. Success in developing value-added capabilities around IoT requires a broad approach that includes expertise in sensing/actuating, connectivity, edge computing, machine learning, networked systems, human-computer interaction, security and privacy. IoT technologies are deployed in different sectors, from agriculture in rural areas to health and wellness, smart home and Smart-X applications in cities.

The IoT is bridging the gap between the virtual, digital and physical worlds by bringing together people, processes, data and things while generating knowledge through IoT applications and platforms. IoT achieves this addressing security, privacy and trust issues across these dimensions in an era where technology, computing power, connectivity, network capacity and the number and types of smart devices are all expected to increase. In this context, IoT is driving the digital transformation as presented in Figure 3.1.

Smart IoT applications with sensing and actuation embedded in "things" are creating smart environments based on hyperconnectivity; the high density of sensing and actuation coverage allows a qualitative change in the

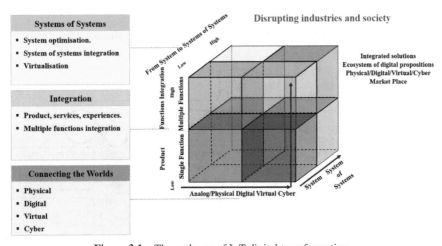

Figure 3.1   The pathway of IoT digital transformation.

way people interact with the intelligent environment cyberspaces, from using appliances at home to caring for patients or elderly persons. The massive deployment of IoT devices creates systems of systems that synergistically interact to form totally new and unpredictable services, providing an unprecedented economic impact that offers multiple opportunities. The potential of the IoT is underexploited; the physical and the intelligent are largely disconnected, requiring a lot of manual effort to find, integrate and use information in a meaningful way. IoT and its advances in intelligent spaces can be categorized with the key technologies at the core of the Internet.

Intelligent spaces are created and enriched by the IoT, in which the traditional distinction between network and device is starting to blur as the functionalities of the two become indistinguishable. With the growing number of IoT deployments, the spectrum of edge devices, short- and long-range radios, infrastructure components from edge computing and cloud storage, as well as networks are increasing in volume, bringing IoT components within reach of a larger pool of potential adopters. In this context, the development of concepts, technologies and solutions to address the perceived security exposure that IoT represents with respect to information technology (IT)/operational technology (OT), is a high priority across several industrial domains, (e.g., manufacturing, automotive, energy, etc.). In Figure 3.2, which will redefine the landscape of business environment.

The IoT as a "global concept" requires a common high-level definition. It has different meanings at different levels of abstraction through the value chain, from lower level semiconductor aspects to service providers. IoT is a paradigm with different visions, and involves multidisciplinary activities.

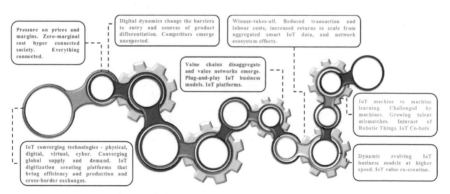

Figure 3.2   The dynamics of IoT digital age.

Considering the wide background and the number of required technologies, from sensing devices, communication subsystems, data aggregation and pre-processing to object instantiation and finally service provision, it is clear that generating an unambiguous definition of the "IoT" is non-trivial.

The IERC is actively involved in ITU-T Study Group 13, which leads the work of the International Telecommunications Union (ITU) on standards for next-generation networks (NGN) and future networks, and has been part of the team which formulated the following definition [10]. "Internet of things (IoT): A global infrastructure for the information society, enabling advanced services by interconnecting (physical and virtual) things based on existing and evolving interoperable information and communication technologies. NOTE 1 – Through the exploitation of identification, data capture, processing and communication capabilities, the IoT makes full use of things to offer services to all kinds of applications, whilst ensuring that security and privacy requirements are fulfilled. NOTE 2 – From a broader perspective, the IoT can be perceived as a vision with technological and societal implications."

The IERC definition [9] states that IoT is: "A dynamic global network infrastructure with self-configuring capabilities based on standard and interoperable communication protocols where physical and virtual 'things' have identities, physical attributes, and virtual personalities and use intelligent interfaces, and are seamlessly integrated into the information network."

3.1.2 IoT Cognitive Transformation

IoT technologies are creating the next generation of smart homes/buildings, smart vehicles and smart manufacturing applications by providing intelligent automation, predictive analytics and proactive intervention. Artificial intelligence (AI) or advanced Machine Learning (ML) is integrated into the different components of IoT architecture layers as part of the complex IoT platforms. These components are composed of many technologies and techniques, (e.g., deep learning, neural networks and Natural Language Processing – NLP). These techniques move beyond traditional rule-based algorithms to create autonomous IoT systems that understand, learn, predict, adapt and operate autonomously and give rise to a spectrum of intelligent implementations, including physical devices, (e.g., robots, autonomous vehicles, consumer electronics) as well as applications and services, (e.g., virtual personal assistants, smart advisors). In this context, the IoT implementations deliver a new class of intelligent applications and things and provide embedded intelligence for a wide range of mesh devices, software platforms and service solutions.

In the IoT world, AI will further enhance the capabilities of concepts such as digital twins, where a dynamic software model is formed of a physical thing or system that relies on sensor data to understand its state, respond to changes, improve operations and add value. Digital twins include a combination of metadata, (e.g., classification, composition and structure), condition or state, (e.g., location and temperature), event data, (e.g., time series) and analytics, (e.g., algorithms and rules) and are used by AI algorithms to model, simulate and predict.

The elements behind the IoT "neuromorphic" structure are illustrated in Figure 3.3.

The cognitive transformation of IoT applications allows the use of optimized solutions for individual applications and the integration of immersive technologies, i.e., virtual reality (VR) and augmented reality (AR); concepts that transform the way individuals and robotic things interact with one another and with IoT platform systems. In this context, VR and AR capabilities are merging with the digital mesh to form a seamless system of intelligent devices capable of orchestrating a flow of information that is delivered to the user as hyper-personalized, hyperconnected and to relevant applications and services. Integration across multiple industrial domains and environments extend immersive applications beyond closed-loop experiences to collaborative cyberspaces of heterogeneous interactive devices and humans. Smart spaces (i.e., rooms, manufacturing floors, and mobility areas) become active with things. Their mesh interconnection will appear and work in conjunction with immersive virtual worlds in a collaborative manner. Cognitive IoT technologies allow embedding intelligence into systems and processes, enabling

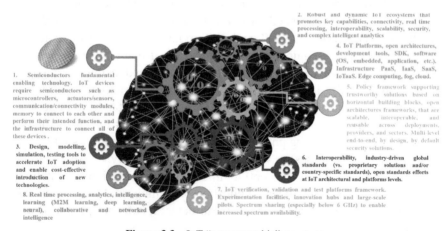

Figure 3.3 IoT "neuromorphic" structure.

the digital mesh to expand the set of endpoints that people and things use to access applications and information or to interact with other people and things. As the device mesh evolves, connection models expand and greater cooperative interaction between devices emerges, creating the foundation for a new continuous and ambient digital experience.

The information exchanged by IoT applications is managed by IoT platforms using cognitive systems with new components addressing the information systems, customer experience, analytics, intelligence and business ecosystems in order to generate new and better services and use cases in the digital business environment.

The cognitive IoT capabilities at the edge integrate the functions of the intelligent digital mesh and related digital technology platforms and application architectures at the cloud level, while increasing the demand for end-to-end security solutions. In addition to the use of established security technologies, it is critical to monitor user and entity behaviour in various IoT scenarios. IoT edge is the new frontier for security solutions creating new vulnerability areas that require new remediation tools and processes that must be embedded into IoT platforms.

The use of artificial intelligence, swarm intelligence and cognitive technologies together with deep learning techniques for optimizing the IoT services provided by IoT applications in smart environments and collaboration spaces, creates new solutions and brings new challenges and opportunities. AI is an increasingly important factor in the development and use of IoT technologies. While focusing on technology it is important to address ethical considerations with respect to deployment and design: ensuring the interpretability of IoT applications and solutions based on AI systems, empowering the consumer, considering responsibility in the deployment of IoT technologies and applications based on AI systems, ensuring accountability and creating a social and economic environment that is formed through the open participation of different stakeholders in the IoT ecosystems.

There are many factors contributing to the challenges faced by stakeholders in the development of IoT technologies based on cognitive capabilities and AI, (i.e., autonomous vehicles, internet of robotic things, digital assistants, etc.), including:

- Decision-making that is based on transparency and "interpretability". When using IoT technologies based on artificial intelligence for performing tasks ranging from self-driving vehicles to managing parking lots or healthcare journals, there is a need for a robust and clear basis for the decisions made by an AI agent. Transparency around algorithmic decisions is in many cases limited by technical secrecy or literacy. Machine learning creates further challenges as the internal decision logic of the model is not understandable even for the developers, and even if the learning algorithm is open and transparent, the model it produces may not be. IoT applications involving autonomous systems need to understand why a self-driving vehicle chooses to take specific actions and need to be able to determine liability in the case of an accident.

- The accuracy and quality of the data that are used by the learning algorithm influence the decisions of an IoT application involving autonomous or robotic vehicles. In these safety-critical and mission-critical applications reliable data are crucial and the use and processing of data from reliable sources is an important element in maintaining confidence and trust in the technology.

- Safety and security are critical for IoT technologies integrated with autonomous systems and AI. Cognitive techniques and AI agents are used to learn about and interact with smart environments, and they must detect unpredictable and harmful behaviour, including indifference to the impact of their actions that can be interpreted as a form of "hacking". In this context, the actions of an AI agent may be limited by how it learns from its environment, how the learning is reinforced and how the exploration/exploitation dilemma is addressed. IoT autonomous systems are exposed to malicious actors trying to manipulate the algorithm by using "adversarial learning" mechanisms to influence the training data for abnormal traffic detection, and this demonstrates that safety and security considerations must be taken into account in the debate around transparency of algorithmic decisions.

- Accountability is another factor that must be considered for IoT autonomous systems based on cognitive and AI technologies where things learn on their own, and humans have less control. Machine learning can create situations that bring into question who is accountable: the producer of the individual thing, the service provider, the fleet manager, the developers/programmers, the collaborative network, etc. The advancement of IoT technologies, requires the issue to be addressed, as flaws in algorithms may result in collateral damages, and there is a need for clarification with regard to liability on the part of the manufacturer, operator and programmer. Cognitive and AI techniques introduce another dimension, as the training data, rather than the algorithm itself, could be the problem.

- The social and economic impacts of IoT technologies based on AI and cognitive solutions are reflected in economic changes through increases in productivity, since robotic things are able to perform new tasks, e.g., self-driving vehicles, networked robotic things or smart assistants to support people in their tasks. This will affect the stakeholders involved in various ways, and create different outcomes for labour markets and society as a whole. IoT autonomous systems improve efficiency and generate cheaper products, create new jobs or increase the demand for certain existing ones, while unskilled and low-paying jobs are more likely to disappear. IoT technologies will have an impact on highly-skilled jobs that rely extensively on routine cognitive tasks. IoT autonomous systems challenge the division of labour on a global scale, and companies may choose to automate their operations locally instead of outsourcing. These developments could increase the digital divide and lead to technological distrust.
- Governance of IoT autonomous systems based on AI and cognitive solutions requires new ways of thinking as these technologies are developed across ecosystems that intersect with topics addressed by the Internet, IoT, AI, robotics governance and policy. Privacy and data laws are experiencing a fundamental paradigm shift as processes are running in parallel with regulations that are adopted or interpreted in different ways. Ensuring a coherent approach in the regulatory space is important, to ensure that the benefits of global IoT technologies, including AI, machine learning, robotics, etc., are realized.

From the point of view of market-based approaches to regulation, all stakeholders should engage to manage the IoT technology's economic and social impact. The social impact of autonomous IoT systems based on cognitive and AI techniques cannot possibly be addressed by governing the technology, and requires efforts to govern the impact of the technology in various applications and domains.

3.2 IoT Strategic Research and Innovation Directions

The IERC brings together EU-funded projects with the aim of defining a common vision for IoT technology and addressing European research challenges. The rationale is to leverage the large potential for IoT-based capabilities and promote the use of the results of existing projects to encourage the convergence of ongoing work; ultimately, the endpoints are to tackle the most important deployment issues, transfer research and knowledge to products and services, and apply these to real IoT applications.

The objectives of IERC are to provide information on research and innovation trends, and to present the state of the art in terms of IoT technology and societal analysis, to apply developments to IoT-funded projects and to market applications and EU policies. The final goal is to test and develop innovative and interoperable IoT solutions in areas of industrial and public interest. The IERC objectives are addressed as an IoT continuum of research, innovation, development, deployment, and adoption.

The IERC launches every year the Strategic Research and Innovation Agenda (SRIA), which is the outcome of discussion involving the projects and stakeholders involved in IERC activities. As such, it brings together the major players of the European landscape to address IoT technology priorities that are essential to the competitiveness of European industry. The SRIA covers the important issues and challenges relating to IoT technology. It provides the vision and roadmap for coordinating and rationalizing current and future research and development efforts in this field, by addressing the different enabling technologies covered by the concept and paradigm of the IoT.

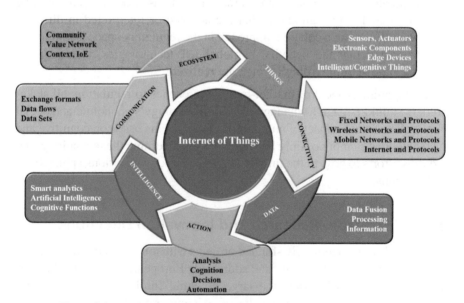

Figure 3.4   IoT components as part of research, innovation, deployment.

Enabled by the activities of the IERC, IoT is bridging physical, digital, virtual, and human spheres through networks, connected processes, and data, and turning them into knowledge and action, so that everything is connected in a large, distributed network. New technological trends bring intelligence and cognition to IoT technologies, protocols, standards, architecture, data acquisition, and analysis, all with a societal, industrial, business, and/or human purpose in mind. The IoT technological trends are presented in the context of integration; hyperconnectivity; digital transformation; and actionable data, information, and knowledge.

IoT developments address highly distributed and hyperconnected IoT applications that use computing platforms, storage, and networking services between edge devices and edge computing and the cloud; these applications drive the growth of new as-a-service business models. Distributed and federated heterogenous IoT platforms at the edge and the cloud as presented in Figure 3.5 require new distributed architectural models to address the future IoT implementations.

The development and deployment of more complex and scalable IoT solutions will result in technological diversification. This will create new challenges for the IoT architecture and open platforms in addressing the complex and cooperative work needed to develop, adopt, and maintain an effective cross-industry technology reference architecture that will allow for true interoperability and ease of deployment. New technological developments in consumers' use of AI-driven IoT opens a new era for IoT; it will be a shift from two-dimensional interfaces for 2D experiences, by using 3D interfaces to generate 3D experiences. In those 3D experiences, things will

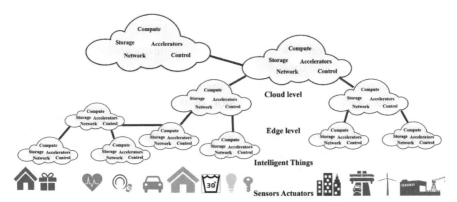

Figure 3.5 Distributed and federated heterogenous IoT platforms at the edge and cloud.

interact with a digital service that takes into account the real-time smart environment and creates a physical result, for example sending a vehicle or robot to a requested location. IoT applications aim to present a single view of data. The convergence of physical, digital, and virtual worlds across multiple channels has created opportunities to measure and influence the product, service and experience beyond traditional value chains, and how stakeholders manage value co-creation.

End-to-end distributed security requires new models and mechanisms to deal with the increased challenges posed by hyperconnectivity. In this context, blockchain technology could be considered the 'missing link' needed to address scalability, privacy, and reliability concerns with respect to IoT technologies and applications. Blockchain technology offers capabilities for tracking a vast number of connected devices; indeed, it can enable coordination and the processing of transactions between devices. The decentralized approach provided by the technology eliminates single points of failure, and thus creates a more resilient device ecosystem. Additionally, the cryptographic algorithms used by blockchain could allow the stronger protection of private consumer data.

The IERC will work to provide a framework that supports the convergence of IoT architecture approaches; it will do so while considering the vertical definition of the architectural layers, end-to-end security, and horizontal interoperability. IoT technology is deployed globally, and supporting the activities of common and unified reference architecture would increase coherence among various IoT platforms. The establishment of a common architectural approach, however, will require a focus on the reference model, specifications, requirements, features, and functionality. These issues will be particularly important in preparing future IoT LSPs, although time schedules might be difficult to synchronize.

The SRIA is developed with the support of a European-led community of interrelated projects and their stakeholders, all of whom are dedicated to the innovation, creation, development, and use of IoT technology.

Since the release of the first version of the SRIA, we have witnessed active research on several IoT topics. On one hand, that research fills several of the gaps originally identified in the SRIA; on the other hand, it creates new challenges and research questions. Recent advances in areas such as cloud computing, cyber-physical systems, robotics, autonomic computing, and social networks have changed even more the scope of convergence in the IoT. The Cluster has the goal of providing an updated document each year that records relevant changes and illustrates emerging challenges.

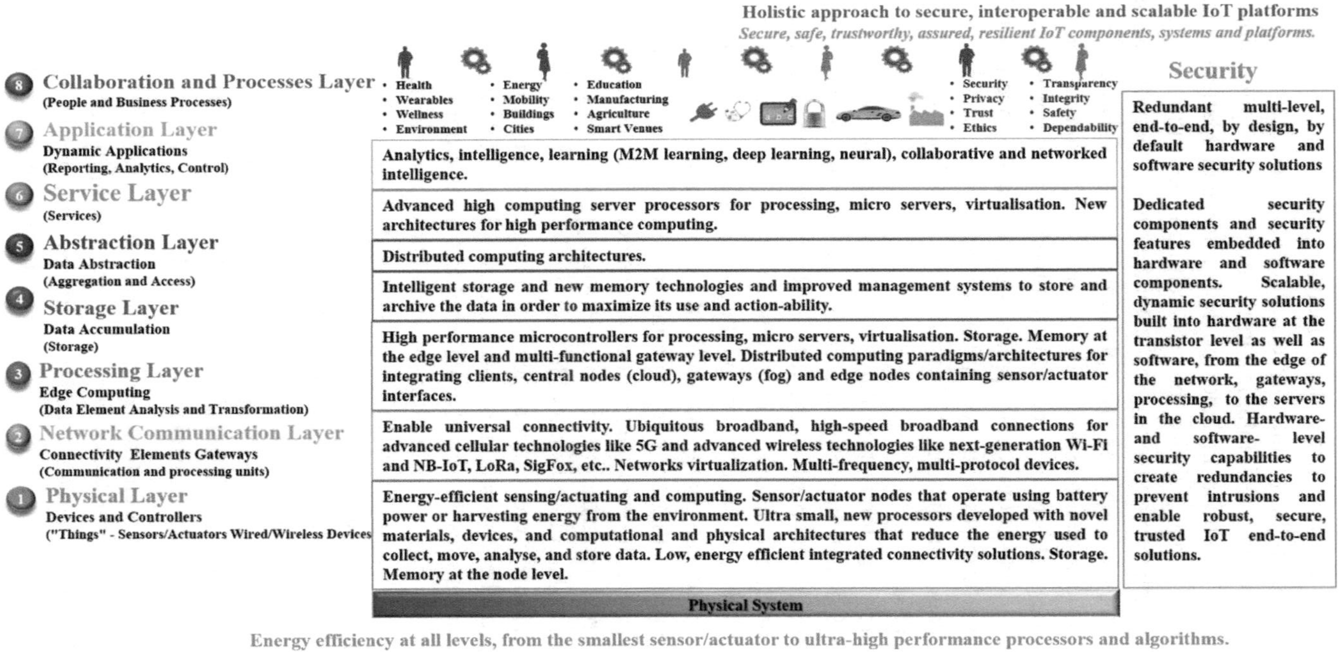

Holistic approach to secure, interoperable and scalable IoT platforms

Secure, safe, trustworthy, assured, resilient IoT components, systems and platforms.

Security

Layer	Description	Security
Collaboration and Processes Layer (People and Business Processes) — Health · Wearables · Wellness · Environment · Energy · Mobility · Buildings · Cities · Education · Manufacturing · Agriculture · Smart Venues · Security · Privacy · Trust · Ethics · Transparency · Integrity · Safety · Dependability	Analytics, intelligence, learning (M2M learning, deep learning, neural), collaborative and networked intelligence.	Redundant multi-level, end-to-end, by design, by default hardware and software security solutions
Application Layer — Dynamic Applications (Reporting, Analytics, Control)	Advanced high computing server processors for processing, micro servers, virtualisation. New architectures for high performance computing.	Dedicated security components and security features embedded into hardware and software components. Scalable, dynamic security solutions built into hardware at the transistor level as well as software, from the edge of the network, gateways, processing, to the servers in the cloud. Hardware-level and software-level security capabilities to create redundancies to prevent intrusions and enable robust, secure, trusted IoT end-to-end solutions.
Service Layer (Services)	Distributed computing architectures.	
Abstraction Layer — Data Abstraction (Aggregation and Access)	Intelligent storage and new memory technologies and improved management systems to store and archive the data in order to maximize its use and action-ability.	
Storage Layer — Data Accumulation (Storage)	High performance microcontrollers for processing, micro servers, virtualisation. Storage. Memory at the edge level and multi-functional gateway level. Distributed computing paradigms/architectures for integrating clients, central nodes (cloud), gateways (fog) and edge nodes containing sensor/actuator interfaces.	
Processing Layer — Edge Computing (Data Element Analysis and Transformation)	Enable universal connectivity. Ubiquitous broadband, high-speed broadband connections for advanced cellular technologies like 5G and advanced wireless technologies like next-generation WI-FI and NB-IoT, LoRa, SigFox, etc.. Networks virtualization. Multi-frequency, multi-protocol devices.	
Network Communication Layer — Connectivity Elements Gateways (Communication and processing units)	Energy-efficient sensing/actuating and computing. Sensor/actuator nodes that operate using battery power or harvesting energy from the environment. Ultra small, new processors developed with novel materials, devices, and computational and physical architectures that reduce the energy used to collect, move, analyse, and store data. Low, energy efficient integrated connectivity solutions. Storage. Memory at the node level.	
Physical Layer — Devices and Controllers ("Things" - Sensors/Actuators Wired/Wireless Devices)		

Physical System

Energy efficiency at all levels, from the smallest sensor/actuator to ultra-high performance processors and algorithms.

Figure 3.6   IoT Research topics addressed at different IoT architectural layers.

Updated releases of this SRIA build incrementally on previous versions [9, 11, 37] and highlight the main research topics associated with the development of IoT-enabling technologies, infrastructure, and applications [1].

The research activities include the IoT European Platforms Initiative (IoT-EPI) program that includes the research and innovation consortia that are working together to deliver an IoT extended into a web of platforms for connected devices and objects. The platforms support smart environments, businesses, services and persons with dynamic and adaptive configuration capabilities. The goal is to overcome the fragmentation of vertically-oriented closed systems, architectures and application areas and move towards open systems and platforms that support multiple applications. IoT-EPI is funded by the European Commission (EC) with EUR 50 million over three years (2016–2018) [16].

The research and innovation items addressed and discussed in the task forces of the IoT–EPI program, the IERC activity chains, and the AIOTI working groups for the basis of the IERC SRIA address the roadmap of IoT technologies and applications; this is done in line with the major economic and societal challenges underscored by the EU 2020 Digital Agenda [36].

The IoT European Large-Scale Pilots Programme [17] includes the innovation consortia that are collaborating to foster the deployment of IoT solutions in Europe through integration of advanced IoT technologies across the value chain, demonstration of multiple IoT applications at scale and in a usage context, and as close as possible to operational conditions.

The programme projects are targeted and goal driven initiatives that propose IoT approaches to specific real-life industrial/societal challenges. They are autonomous entities that involve stakeholders from supply side to demand side, and contain all the technological and innovation elements, the tasks related to the use, application and deployment as well as the development, testing and integration activities.

The scope of IoT European Large-Scale Pilots Programme is to foster the deployment of IoT solutions in Europe through integration of advanced IoT technologies across the value chain, demonstration of multiple IoT applications at scale and in a usage context, and as close as possible to operational conditions. Specific Pilot considerations include:

- Mapping of pilot architecture approaches with validated IoT reference architectures such as IoT-A enabling interoperability across use cases;
- Contribution to strategic activity groups that were defined during the LSP kick-off meeting to foster coherent implementation of the different LSPs.

- Contribution to clustering their results of horizontal nature (interoperability approach, standards, security and privacy approaches, business validation and sustainability, methodologies, metrics, etc.).

IoT European Large-Scale Pilots Programme includes projects addressing the IoT applications based on European relevance, technology readiness and socio-economic interest in Europe. The IoT Large-Scale Pilots projects overview is illustrated in Figure 3.7. IoT European Large-Scale Pilots Programme is funded by the European Commission (EC) with EUR 100 million over three years (2017–2019) [17].

The IERC SRIA is developed incrementally based on its previous versions and focus on the new challenges being identified in the last period.

The updated release of the SRIA highlights the main research topics associated with the development of IoT infrastructures and applications, and it offers an outlook towards 2020 [1].

The timeline of the IERC IoT SRIA covers the current decade (with respect to research), as well as the years that follow (with respect to implementing the research results). As the Internet and its current key applications show, it is anticipated that unexpected trends will emerge that will in turn lead to new and unforeseen development paths.

The IERC has involved experts who work in industry, research, and academia, who provide their vision regarding IoT research challenges, enabling technologies, and key applications that are expected to arise from the current vision for the IoT.

The multidisciplinary nature of IoT technologies and applications reflects in the IoT digital holistic view adapted from [34].

7
CREATE-IoT - CRoss fErtilisation through AlignmenT, synchronisation and Exchanges for IoT: Stimulate collaboration between IoT initiatives, by supporting the development and growth of IoT ecosystems based on open technologies and platforms.

6
U4IoT - User Engagement for Large Scale Pilots in the Internet of Things: Actively engage end-users and citizens to achieve IoT societal acceptance.

5
SYNCHRONICITY - Delivering an IoT enabled Digital Single Market for Europe and Beyond: Single digital city market for Europe.

1
MONICA - Management Of Networked IoT Wearables – Very Large Scale Demonstration of Cultural Societal: Wearable devices containing sensors and actuators for massive scale applications.

2
ACTIVAGE - ACTivating InnoVative IoT smart living environments for AGEing well: Active and healthy ageing.

3
AUTOPILOT - ACTivating InnoVative IoT smart living environments for AGEing well: Automated driving and infrastructure.

4
IoF2020 - Internet of Food and Farm 2020: Strengthen competitiveness of farming and food chains in Europe.

Figure 3.7 IoT European large-scale pilots programme.

The IoT is creating new opportunities and providing competitive advantages for businesses in both current and new markets. IoT-enabling technologies have changed the things that are connected to the Internet, especially with the emergence of tactile Internet and mobile moments (i.e. the moments in which a person or an intelligent device pulls out a device to receive context-aware service in real time). Such technology has been integrated into connected devices, which range from home appliances and automobiles to wearables and virtual assistants.

The IERC SRIA addresses these IoT technologies and covers in a logical manner the vision, technological trends, applications, technological enablers, research agendas, timelines, and priorities, and finally summarizes in two tables future technological developments and research needs.

3.2.1 IoT Research Directions and Challenges

The IoT technologies and applications will bring fundamental changes in individuals' and society's views of how technology and business work in the world. A citizen-centric IoT environment requires tackling new technological trends and challenges. This has an important impact on the research activities that need to be accelerated without compromising the thoroughness, rigorous testing and needed time required for commercialisation.

The integration of billions of "things" in the environment and the functions provided by these things (such as sensing/actuating, interacting and cooperating with each other to enable optimal and efficient services) bring tangible benefits to the environment, economy, citizens and society as a whole and new research challenges. IoT devices involved in IoT applications are very diverse and heterogeneous in terms of resource capabilities, mobility, complexity, communication technologies and lifespan. New research is needed in areas like IoT architecture, communication, naming, discovery, programming models, data and network management, power and energy storage and harvesting, security, trust and privacy. Current Internet approaches are not sufficient to solve these issues, and they need to be revised in order to address the complex requirements imposed by the convergence of industrial, business and consumer IoT. This opens the path for the development of intelligent algorithms, novel network paradigms and new services.

Towards using IoT across industrial sectors, a knowledge-centric network, context awareness, the traffic characterisation, monitoring and optimisation, and the modelling and simulation of large-scale IoT scenarios must be addressed for real-life full-scale deployments, testbeds, prototypes and practical systems.

In Europe, a new dynamic and connected engine for research and innovation is needed in the area of IoT in order to maintain Europe's global edge in IoT research and its innovative spirit and generate new jobs and sustainable economic growth. In this context, an overview of IoT research topics for the coming years is presented below.

A hyperconnected society is converging with a consumer-industrial-business Internet that is based on hyperconnected IoT environments. The latter require new IoT systems architectures that are integrated with network architecture (a knowledge-centric network for IoT), a system design and horizontal interoperable platforms that manage things that are digital, automated and connected, functioning in real time, having remote access and being controlled based on Internet-enabled tools.

Research is not disconnected of development. Thus, the IoT research topics should address technologies that bring benefits, value, context and efficient implementation in different use cases and examples across various applications and industries. The value cycle and the areas targeted by the research activities are presented in Figure 3.8.

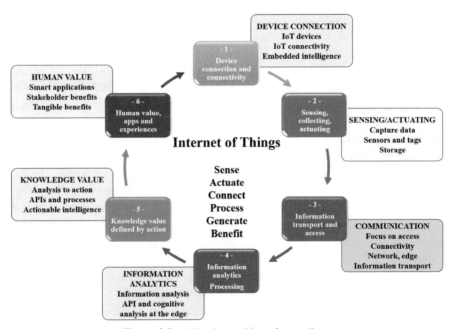

Figure 3.8 IoT value and benefit paradigm.

The shift toward contextual computing, where the intelligent nodes can sense the objective and subjective aspects of a given situation, will augment the ability of edge intelligent "things" to perceive and act in the moment, based on where they are, who they are with, and accumulated experiences. The use of contextual computing in IoT space by combinations of hardware, software, networks, and services that use deep understanding of the intelligent "things" to create costumed, relevant actions that the "things" extend the development of IoT platforms based on new distributed architectures.

The Contextual Internet of Things is the integration of IoT with parallel and opportunistic computing capabilities, neuromorphic and contextual computing (combinations of hardware, software, networks and services) for creating new user experiences and generating tasks on the fly (such as opportunistic IoT applications using data sharing, forming opportunistic networks, on-demand community contextual formation, etc.). Research addressing the context awareness of IoT should include optimal solutions to create and facilitate decentralised opportunistic interactions among humans, IoT networks and the participatory mobile machines. Research should focus on the field of cross-sectorial IoT applications that anticipate human and machine behaviours and human emotions, absorb the social graph, interpret intentions, and provide guidance and support.

The Tactile Internet of Things is based on human-centric sensing/actuating, augmented reality and new IoT network capabilities, including the dynamic mobility of the IoT spatiotemporal systems and data management (personal data, which is consumer-driven, and process data, which is enterprise-driven in a pervasive way). Augmented reality includes 3D visualisation, software robots virtually embedded in things and back-end data systems that enable real-time info and actions. Applications and web browsers are the preferred modes of communication between an IoT device and a smartphone and are challenged by a number of trends and emerging technologies. Messaging platforms for things and developments beyond application program interfaces (APIs) for virtual robots and virtual personal assistants (VPAs) are integrated with things for the post-app era that integrate algorithms at the edge.

The Internet of Mobile Things (IoMT), the Internet of Autonomous Things (IoAT) and the Internet of Robotic Things (IoRT) require research into the area of seamless platform integration, context-based cognitive network integration, new mobile sensor–actuator network paradigms, things identification (addressing and naming in IoT) and dynamic-things discoverability. Research is needed on programmability and communication of multiple heterogeneous mobile, autonomous and robotic things for cooperation, coordination, configuration, exchange of information, security, safety and protection. In addition, research should focus on IoT heterogeneous parallel processing and communication and dynamic systems based on parallelism and concurrency, as well as dynamic maintainability, self-healing and self-repair of resources, changing the resource state, (re-)configuration and context-based IoT systems for service implementation and integration with IoT network service composition.

IoT dynamic collaborative ecosystems are the extension beyond artificial intelligence, where every mobile thing in an IoT application is able to store and analyse its own usage data and then communicate that data smartly to other connected things and make collaborative decisions. When there is a collective networked artificial intelligence and IoT dynamic collaborative ecosystem, the things have the ability to sense, interpret, control, actuate, communicate and negotiate. Networked collaborative artificial intelligence uses natural-language processing and integrated bots (software robots) to interact with users based on deep-learning pattern recognition (vison, speech, smell, sound, etc.), convolutional neural networks and brain-inspired neuromorphic algorithms for parallel processing and communication. This requires developments in the area of dynamic and mobile machine-to-machine learning (beyond basic machine learning) and real-time coordination among mobile-sensing and actuation platforms for coordinated planning. The integration of IoT operating systems and distributed event-stream processing for real-time data analysis is based on distributed stream-computing platforms.

Research onto IoT swarm-based cognition, intelligence and continuous active learning, could lead to the development of IoT programming models through digitisation and automation of the multitudes of heterogeneous things.

Research is needed on IoT horizontal platform integration for providing edge device control and operations, communications, device monitoring and management, security, firmware updates, IoT data acquisition, transformation and management, IoT application development, event-driven logic, application programming, visualisation, analytics and adapters to connect to enterprise systems. Research should also focus on IoT virtual space, mapping and mobility prediction, and virtual deployment for optimising the kinds of mobile things with sensing/actuating capabilities to install, which protocols to use, which types of IoT platforms can send messages directly to each other and which messages need to be routed through gateways or other IoT platforms. Research is also needed on dynamic sensor–actuator fusion and virtual sensing/actuating.

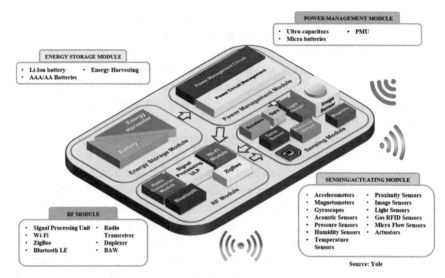

Figure 3.9 IoT sensors/actuators map [12].

Research in the area of sensors/actuators and electronic components that need to integrate multiple function as presented in Figure 3.9.

IoT devices require integrated electronic component solutions that contain sensors/actuators, processing and communication capabilities. These IoT devices make sensing ubiquitous at a very low cost, resulting in extremely strong price pressure on electronic component manufacturers. The research and development in the area of electronic components covers the IoT layered architecture as presented in Figure 3.10.

Additionally, IoT lacks solutions for dynamic context, traffic characterisation- and location-based data processing, storage, processing, virtualisation and visualisation for mobile-edge computing, analytics at the edge (device and gateway level) considering optimal data capture, communication, storage and representation. Moreover, additional work needs to be done in the area of mobile edge-distributed micro IoT clouds based on mobility patterns where data is sent from the same mobile thing to multiple micro IoT clouds. The data needs to be kept synchronised for the purpose of later retrieval and analysis. Research also should focus on how this representation can be extended to data sent from multiple related mobile things.

A context-based end-to-end security framework for heterogeneous devices should be explored for various environments (e.g., operational and information technology security convergence) and applications. For example,

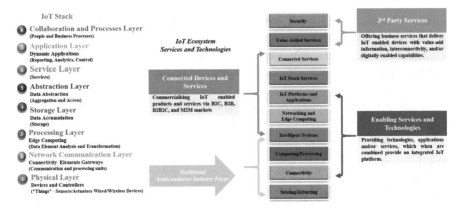

Figure 3.10 IoT electronic devices across the architecture layers.

there is a need for protecting IoT devices and platforms from information cyberattacks and physical tampering by encrypting the communications, as well as addressing new challenges, such as impersonating "things" or denial-of-sleep attacks for batteries. The security framework should be built on real-time business processes and include methods for protecting personal safety and privacy. New artificial intelligence IoT algorithms could be combined with machine-to-machine learning and swarm intelligence to provide new platforms that can identify cyberattacks. Blockchain technology offers capabilities for tracking a vast number of connected devices. It can enable coordination and processing of transactions between devices. The decentralized approach provided by the technology eliminates single points of failure, and thus creates a more resilient device ecosystem.

Data protection in a future IoT landscape with millions of devices continuously monitoring the everyday lives of people is quite challenging. Various attempts have been performed for creating IoT architectures under the concept of privacy by design [30], but still research should be done on creating strong privacy enhancing techniques at the edge, enabling users to have full control over their data in a dynamic way. Research in this area has also to follow the new regulation for data protection of the EU [29].

Heterogeneous networks that combine diverse technical features and low operational cost for various IoT applications should be examined. They can be a mix of short and wide-area networks, offering combined coverage with both high and low bandwidth, achieving good battery life, utilizing lightweight hardware, requiring low operating cost, having high-connection density. When applications request it, the heterogeneous networks should be able to offer high bandwidth, low-latency, high-data rates and a large volume of data, especially in critical applications.

Standardisation and solutions are needed for designing products to support multiple IoT standards or ecosystems and research on new standards and related APIs.

Summarizing, although huge efforts have been made within the IERC community for the design and development of IoT technologies, the always changing IoT landscape and the introduction of new requirements and technologies creates new challenges or raises the need to revisit existing well-acknowledged solutions. Thus, below we list the main open research challenges for the future of IoT:

- IoT architectures considering the requirements of distributed intelligence at the edge, cognition, artificial intelligence, context awareness, tactile applications, heterogeneous devices, end-to-end security, privacy and reliability.
- IoT systems architectures integrated with network architecture forming a knowledge-centric network for IoT.
- Intelligence and context awareness at the IoT edge, using advanced distributed predictive analytics.
- IoT applications that anticipate human and machine behaviours for social support.
- Tactile Internet of Things applications and supportive technologies.
- Augmented reality and virtual reality IoT applications.
- Autonomics in IoT towards the Internet of Autonomous Things.
- Inclusion of robotics in the IoT towards the Internet of Robotic Things.
- Artificial intelligence and machine learning mechanisms for automating IoT processes.
- Distributed IoT systems using securely interconnected and synchronized mobile edge IoT clouds.
- Stronger distributed and end-to-end holistic security solutions for IoT, addressing also key aspects of remotely controlling IoT devices for launching DDoS attacks.
- Stronger privacy solutions, considering the new General Data Protection Regulation (GDPR) [29] for protecting the users' personal data from unauthorized access, employing protective measures (such as PETs) as closer to the user as possible.
- Cross-layer optimization of networking, analytics, security, communication and intelligence.

- IoT-specific heterogeneous networking technologies that consider the diverse requirements of IoT applications, mobile IoT devices, delay tolerant networks, energy consumption, bidirectional communication interfaces that dynamically change characteristics to adapt to application needs, dynamic spectrum access for wireless devices, and multi-radio IoT devices.
- Adaptation of software defined radio and software defined networking technologies in the IoT.

3.3 IoT Smart Environments and Applications

The IoT applications are addressing the societal needs. However, the advancements to enabling technologies such as nanoelectronics and cyber-physical systems continue to be challenged by a variety of technical (i.e., scientific and engineering), institutional, and economical issues.

IoT technologies and applications are driving digital transformation through gathering massive amount of data, rapid deployment of decisions, predictive maintenance and advanced diagnostics, AI and robotic things used in different applications and domains. The IoT applications are expanding from addressing one industrial sector to develop solutions across sectors. Figure 3.11 illustrate the connections between various domains with stronger links when developers are likely to target more verticals.

3.3.1 IoT Use Cases and Applications

As part of the IERC vision, "the major objectives for IoT are the creation of smart environments/spaces and self-aware things (for example: smart transport, products, cities, buildings, rural areas, energy, health, living, etc.) for climate, food, energy, mobility, digital society and health applications" [9].

There has been a swift acceleration in the evolution of connected devices, in terms of both scale and scope, and a greater focus on interoperability. Hyperconnectivity is supported by rapid developments in various communication technologies, including Wi-Fi, Bluetooth, low-power Wi-Fi, Wi-Max, Ethernet, long-term evolution (LTE), and Li-Fi (using light as a medium of communication between the different parts of a typical network including sensors). The hyperconnected and wireless 5G future, which will feature billions of interconnected wireless devices, will require new ways of sharing the spectrum dynamically, using dynamic spectrum access solutions (DSA)

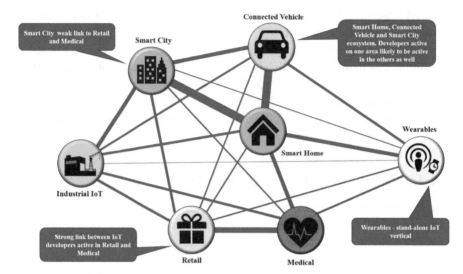

Figure 3.11   IoT connecting people, cities, vehicles, industrial IoT, retail, medical, homes.
Source: VisionMobile 2015.

for low-band, mid-band, and high-band spectrums that will be available for various IoT applications and requirements.

Wireless dedicated IoT communication technologies-such as 3GPP's narrowband NB-IoT, LoRaWAN, or Sigfox-have been deployed in various IoT applications. In this context, standardization and interoperability are critical, as developers, end users, and business decision-makers need to consider more than 36 wireless connectivity solutions and protocols for their applications as presented in Figure 3.12.

The digital economy is based on three pillars: supporting infrastructure (e.g. hardware, software, telecoms, networks), e-business (i.e. processes that an organization conducts over computer-mediated networks) and e-commerce (i.e. the transfer of goods online) [20]. In this new digital environment, IoT software is distributed across cloud services, edge devices, and gateways. New IoT solutions are built on microservices (i.e. application-built modular services, with each component supporting a specific business goal and using a defined interface to communicate with other modules) and containers (i.e. lightweight virtualization) that are deployed and work across this distributed architecture. Machine learning, edge computing, and cloud services, together with AI algorithms, will be used in conjunction with data collected from IoT edge devices.

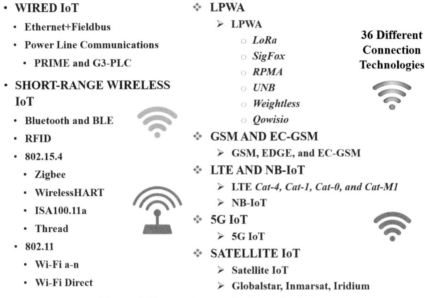

- **WIRED IoT**
  - **Ethernet+Fieldbus**
  - **Power Line Communications**
    - **PRIME and G3-PLC**
- **SHORT-RANGE WIRELESS IoT**
  - **Bluetooth and BLE**
  - **RFID**
  - **802.15.4**
    - **Zigbee**
    - **WirelessHART**
    - **ISA100.11a**
    - **Thread**
  - **802.11**
    - **Wi-Fi a-n**
    - **Wi-Fi Direct**

- ❖ **LPWA**
  - ➤ **LPWA**
    - ○ *LoRa*
    - ○ *SigFox*
    - ○ *RPMA*
    - ○ *UNB*
    - ○ *Weightless*
    - ○ *Qowisio*
- ❖ **GSM AND EC-GSM**
  - ➤ **GSM, EDGE, and EC-GSM**
- ❖ **LTE AND NB-IoT**
  - ➤ **LTE** *Cat-4, Cat-1, Cat-0, and Cat-M1*
  - ➤ **NB-IoT**
- ❖ **5G IoT**
  - ➤ **5G IoT**
- ❖ **SATELLITE IoT**
  - ➤ **Satellite IoT**
  - ➤ **Globalstar, Inmarsat, Iridium**

36 Different Connection Technologies

Figure 3.12   IoT Communication technologies.

3.3.2 Wearables

Wearables are integrating key technologies (e.g. nanoelectronics, organic electronics, sensing, actuating, communication, low power computing, visualisation and embedded software) into intelligent systems to bring new functionalities into clothes, fabrics, patches, watches and other body-mounted devices. The IoT device producers consider that the wearable devices are one of the exciting new markets expected to see the biggest growth over the next few years. The diversity of wearable devices means that the producers will employ 3G or 4G connectivity alongside Wi-Fi to be used for high-speed local connectivity. The drive for low power, leads to many devices being designed for the application accessories. These devices connect via Bluetooth^{TM} LE (Low Energy) or BT (Bluetooth) Smart to a smartphone or tablet to employ its user interface or display, or to process and send data to the Internet and the cloud, linking to services and being part of an IoT application. Wearable technology is enabled by low-power microcontrollers or application processors, low-power wireless chips and sensors, such as MEMS (Micro-Electro-Mechanical Systems) based motion devices and other environmental sensors. Next-generation devices see these devices further miniaturized in highly integrated solutions with ever-smaller batteries to

Wearable Systems Architecture

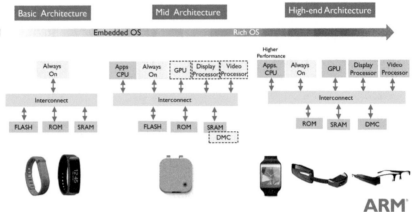

Figure 3.13 Wearables system architecture.

deliver increased functionality in ever-smaller form factors, while high-end products offer increasingly advanced displays and graphics capabilities. In this context, a typical wearables system architecture proposed by companies such as ARM [21] is presented in Figure 3.13.

The global wearable electronics market can be segmented in 5 categories as presented in Figure 3.14 [12]:

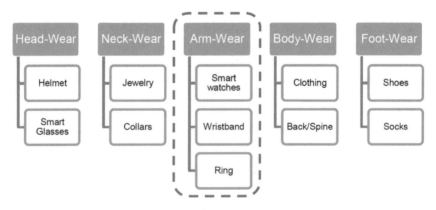

Arm-Wear market is one of the most promising market and many actors are targeting it.

Figure 3.14 Wearable electronic market segmentation [12].

Head-Wear category includes helmet product and vision aid. There's also a category of products for neck-wear, with collars and necklace products that cover up electronics with jewels. Arm-Wear category is the most burgeoning category with multiples devices expected wristband, smartwatches, ring, arm band, etc. Body-Wear products include smart clothing, and devices monitoring back/spine position. The last category concerns foot-wear [12].

CCS Insight expects the wearables market to reach $14 billion by the end of this year and BI Intelligence, Business Insider's research service, expects the wearables market to grow to 162.9 million units by the end of 2020. The healthcare sector is among the top catalysts to push the wearables markets and consumer and professional healthcare trends spur interest in wearable devices. Fitness trackers, are the leading consumer case for wearables as the consumers use wearable devices to record their exercise and health statistics and progress. Hospitals, med-tech companies, pharmaceutical companies, and insurance companies have started to recommend and utilize these devices. One of the major barriers to widespread adoption is accuracy and the manufacturers must ensure that these devices transmit correct data and the users receive accurate progress reports. Privacy concerns are discussed and are not consider as a barrier by early adopters.

Smartwatches offers as well features as fitness bands that could reduce the demand for fitness trackers in the future. The market for wearable computing is expected to grow six-fold, from 46 million units in 2014 to 285 million units in 2018 [35].

The 2016 Gartner Personal Technologies Study surveyed 9,592 online respondents from Australia, the U.S. and the U.K. between June and August 2016, to gain a better understanding of consumers' attitudes toward wearables, particularly their buying behavior for smartwatches, fitness trackers and virtual reality (VR) glasses. According to the survey, smartwatch adoption is still in the early adopter stage (10 percent), while fitness trackers have reached early mainstream (19 percent). Only 8 percent of consumers have used VR glasses/head-mounted displays (excluding cardboard types). The survey found that people typically purchase smartwatches and fitness trackers for their own use, with 34 percent of fitness trackers and 26 percent of smartwatches given as gifts [19].

The innovations are pushing wearable tech into IoT applications for health care, education, smart cities, smart vehicles. The preferred location for wearables has attracted a lot of focus and preferences are shown in Figure 3.15.

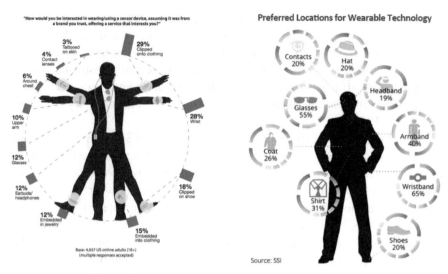

Figure 3.15   Preferred locations for wearable technology.

Source: Google.

The smartwatches will incorporate more sensors, increase functionality and become more autonomous, and they will be untethered from the phone, eschewing Bluetooth connections. The smartwatches will include on board LTE coverage, that allows to call and send texts and connect autonomously to other devices use the device to make payments by swiping the smartwatch at the payment terminal. More computational and communication capabilities more sensors incorporated into smart devices and the smartwatch can act as a hub for other sensors, aggregate and integrate the data from various sensors.

The IoT applications will benefit from the development of wearable technology with integration of virtual and augmented reality features. The virtual information is interfaced to the users using the wearable technology creating ambient AI assistant for coordination and communication and the users will interact with tens different applications to control everything from smart shoes to smart toothbrushes to lightbulbs.

Smart clothing and accessories are integrating seamlessly wearable devices embedded in rings, pendants, sports bras, shoes or clothes including LEDs and colour-changing fabrics. Smart clothing will include many features and different smart solutions are expected on the market in the next years [32, 33]:

- Smart shirt with app, keeping information in 3D showing if too much pressure is put on a certain part of the body, keeping track of your performance, giving information to prevent getting injured while training, with real time feedback
- Health related smart shirt measuring heart rate, breathing rate, sleep monitoring, workout intensity measurements
- Bio sensing silver fibers woven into the shirt
- Clothing to track the number of calories burned
- Clothing to track movement intensity during workout
- Compression fabric that aids in blood circulation and with muscle recovery
- Body monitor sensors – embedded micro sensors throughout the shirt keeping track of temperature, heart beat and heart rate, and the speed and intensity of your workouts
- Shirt able to keep the measured biometrics information by using a small black box woven into the shirt
- Clothing with moisture control and odor control
- Smart shirts can be used in hospitals for monitoring heart beat and breathing in patients
- Baby monitoring – baby garment telling if the baby is sleeping and monitoring the baby's vital signs
- Baby outfit with sensors and a small monitor on it
- Smart socks for baby, monitoring the baby's breath with alert features
- Eco-friendly solar garments as it harnesses the energy of the sun and enables the wearer to charge the owner's phone, music players, and other powered electronic devices
- Adaptive survival clothing that uses moisture and temperature regulation properties of wool to adapt the human body to normal, non-threatening conditions.

The wearable market drivers today are health and fitness smart clothing is used to relay information about fitness and health back to the users. The wearables work well for fitness fans but they still need to reach the everyday consumers since over time, fewer consumers use their wearables daily, which show that the technology isn't becoming habitual or part of the daily routines. In this context, the integration with the IoT digital ecosystem of other products and services is the future trend.

3.3.3 Smart Health, Wellness and Ageing Well

Today, health care stands in a paradigm shift, and new digital solutions require changes in work processes to enable health professionals to spend more time on direct patient contact and treatment. Healthcare and wellness offer unique opportunities for comprehensive IoT implementation. Health care treatments, cost, and availability affect the society and the citizens striving for longer, healthier lives. IoT is an enabler to achieve improved care for patients and providers. It could drive better asset utilization, new revenues, and reduced costs. In addition, it has the potential to change how health care is delivered.

The emergence of Internet of Health (IoH) applications dedicated to citizens health and wellness that spans care, monitoring, diagnostics, medication administration, fitness, etc. will allow the citizens to be more involved with their healthcare. The end-users could access medical records, track the vitals signals with wearable devices, get diagnostic lab tests conducted at home or at the office building, and monitor the health-related habits with Web-based applications on smart mobile devices. Smart Home and welfare technology will merge in integrated services for the benefit of both residents and the municipality. The solutions need to be tailored according to individual needs and evolve as care needs increase. Health information should in future accompany the patient throughout life. IoT systems should be based on patients "and services" needs while confidentiality and privacy are protected. Both the current and future needs for quality-assured information sharing across service levels and business boundaries in the health and care sector, and with other government agencies must implement the new systems.

The IoT technologies offer different solutions for healthcare applications starting from traditional one to wearables and "gadgets" that still need to be develop and tested as listed below:

- Teeth. Toothbrushes that will measure fluoride, remember cavities and discoloration, and notify you of bad breath.
- Eyes. Glasses that will monitor your eyesight and advise correction.
- Hair. Combs that will screen the follicles, report on dandruff density, scan for fungus or lice, and count the hairs (hair loss).
- Bottom. Toilets that will test excrements, both liquid and solid.
- Chest. Airport scanners that will broadcast their results to your phone.
- Body. Clothes that will be intelligent because the fibres will compute, and that will visualize your body language.
- Underbelly. A new field of under wearables that will integrate markers for early detection of cancers or other anomalies.

- Forearms. Shirts that will screen the microbiome on your forearms (40x more than our own cells).
- Neck. Collars that will chemically analyse your sweat.
- Ear. Earphones that will measure your hearing and analyse the emotional level of people you are listening to (sound analysis already allows that!), interesting for total communication (i.e. beyond words and including body language).
- Heart. Pacemakers and stents that will broadcast data to the cardiologist plus ECG
- Nose. Tissues that will examine snot and mucus when you blow your nose.
- Chin. Razors that will plot the surface of the skin looking for acne.
- Lips. Balm that will scan for cold sores.
- Tongue. Tongue scrapers that will screen salivary microbes (the oral microbiome).
- Back. Chairs that will plot your posture and broadcast data for your spine.
- Nails. Nail cutters that will determine the quality of your nails and count the ridges.
- Feet. Step counters.
- Pulse. Heart rate monitors.
- Brain. Headsets that will measure electrical activity in the form of alpha, beta, delta and theta waves.

The World Health Organization (WHO) defines e-Health as: "E-health is the transfer of health resources and health care by electronic means. It encompasses three main areas: The delivery of health information, for health professionals and health consumers, through the Internet and telecommunications; Using the power of IT and e-commerce to improve public health services, e.g. through the education and training of health workers, the use of e-commerce and e-business practices in health systems management. E-health provides a new method for using health resources – such as information, money, and medicines – and in time should help to improve efficient use of these resources. The Internet also provides a new medium for information dissemination, and for interaction and collaboration among institutions, health professionals, health providers and the public."

IoT applications have a market potential for electronic health services and connected telecommunication industry with the possibility of building ecosystems in different application areas.

The smart living environments at home, at work, in public spaces should be based upon integrated systems of a range of IoT-based technologies and services with user-friendly configuration and management of connected technologies for indoors and outdoors. These systems can provide seamless services and handle flexible connectivity while users are switching contexts and moving in their living environments and be integrated with other application domains such as energy, transport, or smart cities. The advanced IoT technologies, using and extending available open service platforms, standardised ontologies and open standardised APIs can offer many of such smart environment developments.

These IoT technologies can propose user-centric multi-disciplinary solutions that take into account the specific requirements for accessibility, usability, cost efficiency, personalisation and adaptation arising from the application requirements. IoT technology allows that a variety of functions are controlled with various sensor, hardware, communication, cloud and analytics and integrated, with the living environments allow people with a range of needs to retain their independence. The IoT technology not only overcomes the inconvenience of distance, but also provides people with greater choice and control over the time and the place for monitoring their condition, increasing convenience and making their conditions more manageable. At the same time, it also reduces some of the pressures on clinics and acute hospitals. IoT could make a significant contribution to the management of several chronic conditions, heart failure, hypertension, asthma, diabetes and can be integrated with other living environments domains such as mobility, home/buildings, energy, lighting, cities.

In this context, the IoT applications need to be included into an integrated IoT framework for active and healthy living and sustainable healthcare as presented in Figure 3.16. This need to be implemented using an IoT architecture model for convergence between social and health services, supporting older people and those with long term conditions to live independently and lead fulfilling lives with the national healthcare architecture.

The IoT distributed architecture is built in a modular manner, designed to logically isolate safety critical and non-safety critical systems elements, provide standard open integration points to collaborating systems components, and to take advantage of the principles of service oriented approaches to systems design. Real time and batched event and health metric data are acquired by highly available, resilient and performant modules. The data provided by these modules is then exposed by a series of web services which provide means by which local and remote staff and systems can manage

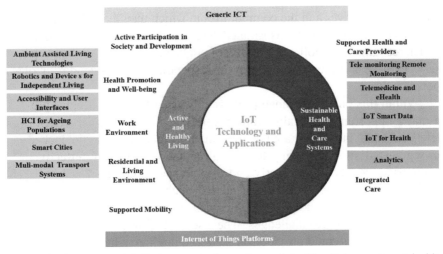

Figure 3.16 Integrated IoT framework for active and healthy living and sustainable healthcare.

the tele monitoring service more effectively. New applications will allow individual user roles to more efficiently deliver tele monitoring services via a single interface. Both applications and services operate in concert on an integrated IoT platform providing data and business logic integration including the services and workflow.

Demographic change, the rising incidence of chronic disease and unmet demand for more personalised care are trends requiring a new, integrated approach to health and social care. Such integration – if brought about in the right manner – has the potential to improve both the quality and the efficiency of care service delivery. Potentially this can be to the benefit of all: beginning with older people in need of care and their family and friends, and including care professionals, service provider organisations, payers and other governance bodies.

There is a need for fundamental shift in the way we think about older people, from dependency and deficit towards independence and well-being. Older people value having choice and control over how they live their lives and interdependence is a central component of older people's well-being. They require comfortable, secure homes, safe neighbourhoods, friendships and opportunities for learning and leisure, the ability to get out and about, an adequate income, good, relevant information and the ability to keep active and healthy. They want to be involved in making decisions about the questions that affect their lives and the communities in which they live. They also want services to be delivered not as isolated elements, but as joined- up provision, which recognises the collective impact of public services on their lives. Public services have a critical role to play in responding to the agenda for older people.

Within this ongoing change process, advanced IoT technologies provide a major opportunity to realise care integration. At the same time, telecare, telehealth and other IoT applications in this field also remain locked up in segregated silos, reflecting the overall situation. Providing effective and appropriate healthcare to elderly and disable people home is a priority and the use of seamless information and IoT technology at home, in public places, in transport, energy and cities can enable healthcare management to mitigate the future challenges. The use of IoT technologies integrated with other sectors could provide complete and intelligent health management services to elderly home, which provides sustainable healthcare service for elderly people. These new solutions make both the elderly life easier and the healthcare process more effective.

Challenges are the integration of software and hardware to give improved performance of the IoT gateway, provided through various IoT platforms, an enhanced data and network security including down the edge device management for software updates and configuration changes. Successful IoT solutions for elderly people must address:

- Ease-of-use considering that many elderly people aren't comfortable with technology or face issues such as diminished vision or arthritis.
- Non-stigmatizing, "invisible" that cannot be visible and used to further identify and isolate elderly people.
- Privacy and security in order to avoid elderly people to be targets of scams and actions to exploit them or becoming more vulnerable, considering their health situation and health conditions.
- Affordability of IoT technology with devices that are low cost and reliable and can be covered on a fixed income.
- Technology that encourage mutual support and motivators.
- Support and foster independence combing wearables, smart mobility, smart home, smart city applications that help elderly people manage their daily lives to increase the chance they could stay in their homes and move in the city independently for longer, an important factor in both reducing hospitalization costs and fostering self-worth.

3.3.4 Smart Buildings and Architecture

Buildings consume 33% of world energy, this figure grows to 53% of world electricity, and it will continue to grow in the future. As a result, buildings have an important weight in regards to the energy challenge. Today, most commercial buildings are having basic infrastructure to its purpose or costs and many building managers lack basic visibility into the infrastructure they are responsible for.

The current lack of dispatchablility is the fundamental disconnect between the current state in which buildings are passive, "sleeping" untapped assets for operators and building owners, and the future state, in which buildings could act as distributed energy assets, functioning as "shock absorbers" for the grid, opening up new value streams for owners and operators, and, in general, playing an essential role in enabling a more efficient, green, and secure energy system. The current system does not exchange energy data information between assets in buildings and between buildings and the grid. It is comprised of legacy distribution management systems (DMS) that make up the backbone of the utility's grid control and optimization systems, installed distributed energy resources (DER), including PV, fuel cells, and combined heat and power systems, and the ultimate customer-side loads (i.e., the buildings and their equipment, appliances, and devices that ultimately "consume" the energy) [25].

Improving life of the occupants implies many aspects including comfort with light, temperature, air quality, having access to services facilitating life inside the building, adapting the behaviour to the needs of the occupants. There is also a direct economic interest to do it as it is recognized that productivity level is connected to the comfort level.

In this context, IoT is already having a significant impact on the commercial real estate (CRE) industry, helping companies move beyond a focus on cost reduction. IoT applications aim to grow margins and enable features such as dramatically more efficient building operations, enhanced tenant relationships, and new revenue generation opportunities [23].

The different ingredients of IoT, connectivity, control, cloud computing, data analytics, can all contribute to make smarter buildings (offices, industrial, residential, tertiary, hotels, hospitals, etc.):

- Connected to the grid ("smart grid ready")
- Connected to the smart city
- Energy efficient while taking care of the comfort of the occupants
- Adaptable to the changing needs of the occupants over time

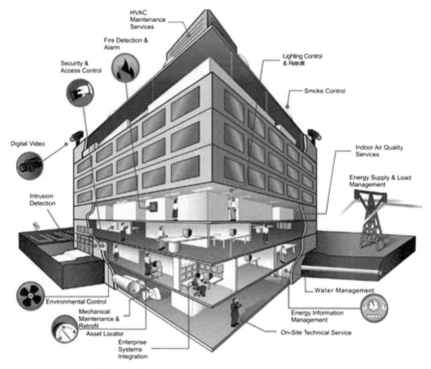

Figure 3.17   Smart building implementation [22].

- Providing services for a better life of the occupants
- Easy to maintain during the whole life cycle at minimal cost

The solutions focus primarily on environmental monitoring, energy management, assisted living, comfort, and convenience. Utilizing the IoT platforms in the houses and buildings, heterogeneous equipment empowers the automation of regular activities. Through transforming things into appliances' data which are thoroughly linked by applying the Internet can implement services through web interfaces. The solutions are based on open platforms that employ a network of intelligent sensors to provide information about the state of the home. These sensors monitor systems such as energy generation and metering; heating, ventilation, and air-conditioning (HVAC); lighting; security; and environmental key performance indicators.

The uses of the IoT connected with automate building maintenance activities, primarily aiming to realize the benefits of low-hanging fruit such as cost savings and operational efficiency through improved energy management and reduced personnel costs. The approaches are focusing on developing connected Building Management Systems (BMS) that are progressively more connected and integrated. Different approaches to develop connected BMS are presented in [23] and are categorised as:

- Individual BMS: CRE owners install BMS on a piecemeal basis to automate individual tasks such as elevator or lighting control;5 not surprisingly, owners then must collect and aggregate data from various places.
- Partially integrated BMS: CRE companies are using partially integrated BMS, combining automation of a few activities with a common focus, such as energy management systems. Compared with individual BMS, these systems are more integrated, require less manual intervention, and enable faster decision making. More importantly, CRE owners use these systems to enhance tenant and end-client experience through sustainability initiatives (including to support LEED and other green building certification standards), open Wi-Fi access, and so forth.
- Fully integrated, IoT-enabled BMS: IoT-enabled systems fully integrated BMS, allowing higher-order cost, productivity, and revenue benefits with a deep customer and data focus. It can leverage one infrastructure to operate all building management solutions and require minimal to no manual involvement. Internet protocol or IP-enabled devices can facilitate intelligent decision making by automating point decisions and enhancing strategic insights; this allows data to automatically flow all the way around the Information Value Loop without manual interaction, enabling quick action on the data and creating new value for CRE companies.

In order to improve the technology integration and interoperability the following approaches should be considered [23]:

- Develop advanced mobile computing capabilities: CRE companies will likely benefit from developing a flexible mobile application platform that can integrate new IoT information tracking and capture requirements.
- Use appropriate integration software and platforms: Owners of existing buildings can consider buying specialist software solutions that integrate siloed and disparate building systems and improve interoperability. Likewise, owners of new buildings should consider adopting the latest integrated IoT platforms.
- Use common standards and protocols: Gradual consolidation of different BMS protocols will help develop benchmarks that facilitate full use of

IoT technology. OASIS Open Building Information Exchange is one global industry-wide effort aiming to define standard web protocols for communication between various BMS. Ultimately, players must agree on benchmarks to increase interoperability even among systems used by different industries.

The Internet of Building (IoB) and Building Internet of Things (BIoT) concepts integrates the information from multiple intelligent building management systems and optimise the behaviour of individual buildings as part of a larger information system. The value in IoB is as much in the edge devices and the data collected, exchanged and processed. Collecting, exchanging and processing data from building services and equipment provides a granular view of how each building is performing, allowing the development of building systems that collect, store and analyse data at the edge and in the cloud, providing better operational efficiency and integration with IoT platforms and applications across various sectors.

The implementation of the IoB concept in residential building environments require to integrate the IoT gateway architecture into a flexible data platform that is able to run parallel metering and non-metering applications

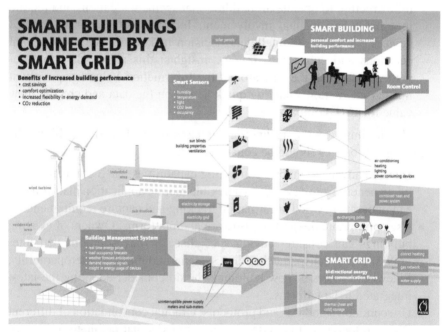

Figure 3.18 Smart Building connected by a Smart Grid [24].

in real-time. An example of such possible implementation for DC drids in residential buildings is shown in Figure 3.19. The solution is able to generate large volumes of granular energy data and behavioural information, and it is able to process these energy and behavioural data locally. It is likely that for the smart energy domain, this solution will evolve into reference residential gateways designs that combines energy management services with other vertical applications in a heterogeneous networking environment.

The future energy economy based on IoT technology need to include open, interoperable transaction-based platforms that facilitates physical transactions of energy, energy-related services, and the financial settlements associated with these transactions and integration with the smart grid (see Figure 3.18), smart city, lighting, mobility applications.

3.3.5 Smart Energy

The energy supply will be largely based on various renewable resources and this source of energy will influence the energy consumption behaviour, demanding an intelligent and flexible electrical grid which is able to react to power fluctuations by controlling electrical energy sources (generation, storage) and sinks (load, storage) and by suitable reconfiguration. The functions

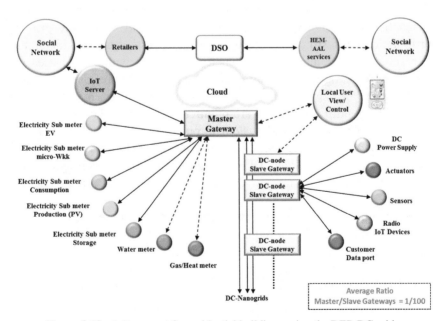

Figure 3.19   IoT concept for residential buildings using the DER DC grid.

are based on networked intelligent devices (appliances, micro-generation equipment, infrastructure, consumer products) and grid infrastructure elements, largely based on IoT concepts.

IoT is expected to facilitate the deployment of new smart energy apps within energy stakeholders' ICT systems (generation and retail companies, grid and market operators, new load aggregators) bringing new options for real-time control strategies across energy asset portfolios for faster reactions to power fluctuations. These new technologies combine both centralised and decentralised approaches integrating all energy generation (generation, storage) and load (demand responsive loads in residential, buildings and industries as well as storage and electrical vehicles) through interconnected real-time energy markets. IoT should also improve the management of asset performance through more accurate estimations of asset health conditions and deployment of fact based preventive maintenance.

These new smart energy apps will largely be based on the networking of IoT intelligent devices embedded within Distributed Energy Resources (DER) spread across the energy system such as consumer appliances, heating and air conditioning, lighting, distributed generation and associated inverters, grid edge and feeder automation, storage and EV charging infrastructures. While energy systems have historically been controlled through single central dispatch strategies with limited information on smart grid edge and consumers behaviours, energy systems are characterized by rapidly growing portfolios of DER structured through several layers of control hierarchies interconnecting the main Grid down to microgrids within industries and communities, nanogrids at building level and picogrids at residential scale.

Most of DER have diffused within end-user premises, new transactive energy (TE) control approaches are required to facilitate their coordination at various scales of the Grid system through real-time pricing strategies. The aggregators and energy supply companies have started to develop new flexibility offers to facilitate DER coordination virtually through ad hoc virtual power plants raising new connectivity, security and data ownership challenges.

Meanwhile climate change has also recently exposed grids to new extreme weather conditions requiring to reconsider grid physical and ICT architectures to allow self-healing during significant disasters while taking advantage of Distributed Generation and storage to island critical grid areas (hospital, large public campus) and maintain safe city areas during emergency weather conditions.

By 2030, the future utility value chain will have transformed significantly. Navigant Research argues that current distribution network operators will have transformed into distribution service orchestrators; they will be responsible for far more than just network operations. Likewise, the current energy supply business – already transitioning to a service-based model – will be fully transformed into an energy service provider (ESP) model. Companies will offer end-to-end energy services that have little in common with today's volume-based approach to revenue generation. The resulting new business models will require new IT infrastructure that relies heavily on the analysis of huge volumes of data. Distribution orchestration platforms will rely on the integration of existing advanced distribution management systems (ADMSs) and DER management software, as well as the incorporation of a market pricing mechanism to reflect the changing value of millions of connected endpoints throughout the day. ESPs will rely on TE platforms that enable prosumers to sell their power into the market, incorporate customer portals to provide in-depth account details, and provide billing and settlement functionality [26].

The high number of distributed small and medium sized energy sources and power plants can be combined virtually ad hoc to virtual power plants. Using this concept, areas of the grid can be isolated from the central grid and supplied from within by internal energy sources such as photovoltaics on the roofs, block heat and power plants or energy storages of a residential area. Microgrids and Nanogrids either islanded or grid-connected are compounded by several agents of different nature. Consumers, producers and prosumers aim to achieve specific local goals such as reliability, diversification of energy sources, low carbon emission and cost reduction. Small-scale storage is offering flexibility to the electric power system, which can contribute to increase grid security and stability, modifying the electricity generation and load patterns in grid nodes. At the same time, this means an increase in reliability, power quality and renewable energy penetration. Due to the high penetration of renewables, energy storage usage and the wide variation of different resources, electrical grid environment is getting more and more complex. In the energy domain, the proliferation of microgrids for local control of energy sources, integration of renewables and energy storage units requires the integration of communication gateways and distributed cooperative control for bidirectional energy flow. In this context, energy flows will be managed similarly to internet data packets across grid nodes, which autonomously decide the best pathway minimising energy system dispatch costs while guaranteeing its best resiliency.

This is based on the development of Internet of Energy (IoE) concept as network infrastructure based on standard and interoperable communication nodes that allow the end to end real time balance between the local and the central generation, responsive demand and storage. It will allow units of energy to be transferred peer to peer when and where it is needed. For these applications, the IoT gateway is integral component of the micro inverter system and operates between the micro inverters and energy brokers systems and the web-based monitoring and analysis software. A conceptual representation of an integrated energy system based on DER, microgrids where the communication gateways play a key role is presented in Figure 3.20.

The requirements for the IoT technologies and gateways in the energy sector has to be seen in relation with the development of the virtual power plants and novel virtual microgrid architectures. These requirements include microgrid communication infrastructures for microgrid applications such as inter substation communication, substation to building communication, and distributed energy resources. The microgrid information management systems for distribution automation demand side scheduling, distributed generation/control, real time unit embedded system for substations, medium voltage measurement, monitoring and communication system.

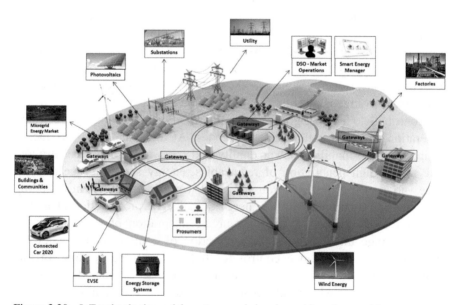

Figure 3.20　IoT technologies and the gateway role in microgrids and nanogrids management systems.

These IoT technologies should combine both centralised and decentralised approaches integrating all energy generation (generation, storage) and load (demand responsive loads in residential, buildings and industries as well as storage and electrical vehicles) through interconnected real-time energy markets. IoT should also improve the management of asset performance through more accurate estimations of asset health conditions and deployment of fact based preventive maintenance.

The IoT technologies deployment will have a significant impact on the energy industry with a shift in business models from energy supply-based to service-based models.

The service-based model of the 2030s means network operators' primary focus is now on end customers and networks are far more dynamic, volatile, and unpredictable [26]:

- The rise of the prosumer means that power flows have become two-way.
- Self-consumption by PV and electricity storage owners significantly changes load curves to evening peaks when solar PV is no longer generating power.
- New, power-hungry appliances such as EVs and heat pumps place significant new demands on network capacity.
- The aggregation of DER into virtual power plants creates dispatchable power connected directly to low voltage networks.
- TE systems encourage rapid switching between the export and import of power from many premises throughout the day.
- DER aggregation and management become a critical component of a distribution service orchestrator's role. Network volatility and dynamism require more active management of low voltage networks, particularly managing the peak consumption periods when customers shift from self-generated consumption to grid-sourced electricity.

The 2030 energy landscape, presented in [26], has the customer in the center of the Energy Cloud (Figure 3.21) with the following characteristics:

- The Energy Cloud is a mature set of technologies. Ubiquitous solar PV and storage create a customer-centric energy value chain where customers' consumption is largely met by self-generated electricity.
- Utility-scale and distributed renewables account for 50%–100% of generation; distributed energy resources (DER) uptake is widespread, accounting for most new build capacity.
- High penetration rates of EVs put a strain on network capacity, which is managed using pricing signals and automatic demand response (DR).

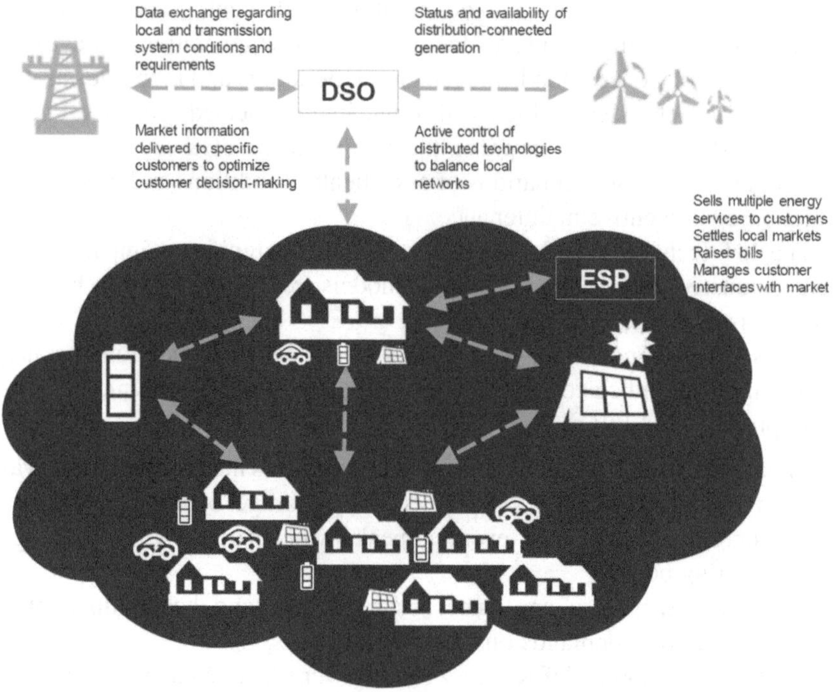

Data exchange regarding local and transmission system conditions and requirements

Status and availability of distribution-connected generation

DSO

Market information delivered to specific customers to optimize customer decision-making

Active control of distributed technologies to balance local networks

Sells multiple energy services to customers
Settles local markets
Raises bills
Manages customer interfaces with market

ESP

Figure 3.21   Energy cloud 2030 [24].

- Data is as valuable a commodity as electrons. While in 2017 the industry struggled to maximize the value of enterprise data, in 2030 the energy supply chain is fully digitized and its efficient operation is heavily based on analytics-based automation. This automation relies on the huge volumes of data created by technologies within the Energy Cloud.
- The industry has undergone significant digital transformation. Data and artificial intelligence (AI)-based algorithms become important competitive differentiators. Data offers visibility into each prosumer's electricity exports and imports, providing the fundamental basis of the transitive energy market. This data also allows the newly formed distribution service orchestrators to actively manage the dynamic and volatile distribution networks, either through pricing signals or by actively interrupting the power supply.
- Utility business models have transformed from supply- to service-based. Rather than focus purely on the delivery of grid-sourced power, energy service providers (ESPs) offer individualized products and services to

suit their customers' specific needs. These services will include DER sales, maintenance, and aggregation; DR; energy efficiency initiatives; flexible, time-of-use charging; and TE platforms.

- Markets are far more competitive in 2030 compared to 2017. The convergence of the old regulated supply business model and deregulated service-based model creates opportunities for new entrants. Many new service providers have entered the market, exploiting the new value streams from decentralized electricity.
- The smart grid of 2017 has transitioned to a neural grid. The new grid is nearly autonomous and self-healing, leveraging innovations in AI and cyber-physical systems (e.g., IoT, self-driving EVs, and the smart grid).
- Distribution operators have evolved into distribution service orchestrators to manage this neural grid. Advanced platforms incorporate advanced distribution management systems (ADMSs), DER management systems (DERMSs), and pricing signals to manage the more volatile and dynamic grids.
- Two separate, yet complementary technology platforms underpin the market. TE and distribution orchestration platforms enable prosumers to sell self-generated power on open markets and manage the highly volatile and dynamic distribution networks.
- In 2030, prosumers trade their self-generated power on the open market. This is a dramatic change from relying on the subsidies or net metering that supported residential solar PV in 2017. Electricity is bought and sold at market rates and revenue from a TE platform alone totals $6 billion per year. To bring this into the perspective of other disruptive innovators, TE revenue in 2030 is 4 times the size of Uber's 2015 revenue.

3.3.6 Smart Mobility and Transport

Consumer preferences, tightening regulation, and technological breakthroughs add up to a fundamental shift in individual mobility behaviour. Individuals increasingly use multiple modes of transportation to complete their journey, and goods and services are increasingly delivered to (rather than fetched by) consumers. As a result, the traditional business model of car sales will be complemented by a range of diverse on-demand mobility solutions, especially in dense urban environments that proactively discourage private car use. Consumers today use their cars as "all-purpose" vehicles (Figure 3.22), no matter if commuting alone to work or taking the whole family to the beach. In the future, they may want the flexibility to choose the

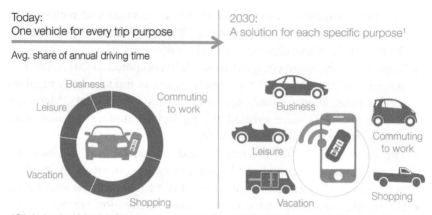

Figure 3.22   Mobility patterns [15].

best solution for a specific purpose, on demand and via their smartphones. We can already observe significant, early signs that the importance of private car ownership is declining and shared mobility is increasing [15].

The connection of vehicles to the Internet offers new possibilities and applications which bring new functionalities to the individuals and/or the making of transport easier and safer. New mobile ecosystems based on trust, security and convenience to mobile/contactless services and transportation applications are created and the developments such as Internet of Vehicles (IoV) [38] are connected with Internet of Energy (IoE) for providing services in an increasingly electrified mobility industry.

Representing human behaviour in the design, development, and operation of cyber-physical systems in autonomous vehicles is a challenge. Incorporating human-in-the-loop considerations is critical to safety, dependability, and predictability. There is currently limited understanding of how driver behaviour will be affected by adaptive traffic control cyber-physical systems.

Self-driving vehicles today are evolving and the vehicles are equipped with technology that can be used to help understand the environment around them by detecting pedestrians, traffic lights, collisions, drowsy drivers, and road lane markings. Those tasks initially are more the sort of thing that would help a driver in unusual circumstances rather than take over full time.

Technical elements of such systems are smart vehicle on-board units which acquire information from the user (e.g. position, destination and schedule) and from on board systems (e.g. vehicle status, position, energy usage profile, driving profile). They interact with external systems (e.g. traffic control systems, parking management, vehicle sharing managements, electric vehicle charging infrastructure).

In the field of connected autonomous vehicles IoT and sensing technology replace human senses and advances are needed in areas such as [27]:

- Vehicle's location and environment: As there would no longer be active human input for vehicle functions, highly precise and real-time information of a vehicle's location and its surrounding environment will be required (e.g., road signs, pedestrian traffic, curbs, obstacles, traffic rules).
- Prediction and decision algorithms: Advanced concepts based on Artificial Neural Networks (unsupervised/deep learning, machine learning) will be needed to create systems to detect, predict and react to the behaviour of other road users, including other vehicles, pedestrians and animals.
- High accuracy, real time maps: Detailed and complete maps must be available to provide additional and redundant information for the environmental models that vehicles will use for path and trajectory planning.
- Vehicle driver interface: A self-adapting interface with smooth transition of control to/from the driver, mechanisms to keep the driver alert and a flawless ride experience will be instrumental in winning consumer confidence.

Successful deployment of safe and autonomous vehicles (SAE[1] international level 5, full automation) in different use case scenarios, using local and distributed information and intelligence is based on real-time reliable platforms managing mixed mission and safety critical vehicle services, advanced sensors/actuators, navigation and cognitive decision-making technology, interconnectivity between vehicles (V2V) and vehicle to infrastructure (V2I) communication. There is a need to demonstrate in real-life environments (i.e. highways, congested urban environment, and/or dedicated lanes), mixing autonomous connected vehicles and legacy vehicles the functionalities in order to evaluate and demonstrate dependability, robustness and resilience of the technology over longer period of time and under a large variety of conditions.

The evolutions in the global automotive industry are monitored in order to identify the factors that are driving the change in the automotive ecosystem,

---

[1]Society of Automotive Engineers, J3016 standard.

the move to new business models such as mobility-as-a-service and the best conditions and the technologies that are supporting the digital transformation. The automotive industry has followed a very linear development path in more than 100 years. The parallel emergence of four megatrends in the last 2 years mobility, automated driving, digital experience and electrification, the industry will be reshaped in the next 10 to 15 years.

The Radar presented in [28] analyzes the transformation via 25 selected indicators in five dimensions: customer interest (e.g. via >10,000 end user interviews), regulation, technology, infrastructure and industry activity.

Figure 3.23 presents the automotive disruption radar globally that indicates the customers' interest in autonomous vehicles and their acceptance of electrical vehicles as an alternative, but a low interest in vehicle to vehicle communication.

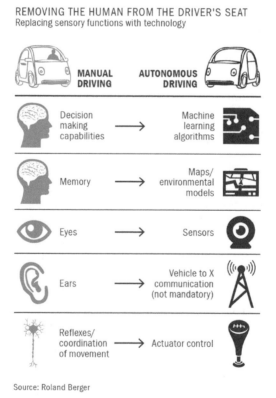

Figure 3.23   Replacing sensory functions with technology [27].

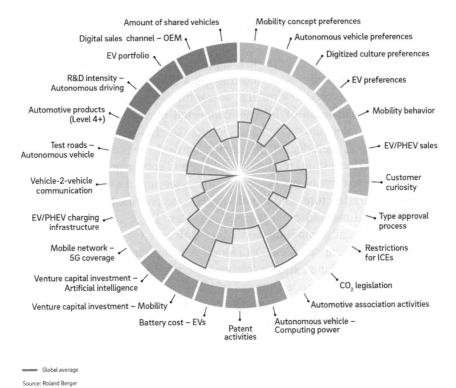

Figure 3.24 Automotive disruption radar globally [28].

3.4 IoT and Related Future Internet Technologies

3.4.1 Edge Computing

The use of intelligent edge devices require to reduce the amount of data sent to the cloud through quality filtering and aggregation and the integration of more functions into intelligent devices and gateways closer to the edge reduces latency. By moving the intelligence to the edge, the local devices can generate value when there are challenges related to transferring data to the cloud. This will allow as well for protocol consolidation by controlling the various ways devices can communicate with each other. There are different edge computing paradigms, such as transparent computing and fog computing. The fog computing is focusing on resource allocation in the service level, while transparent computing concentrates on logically splitting the software stack (including OS) from the underlying hardware platform to provide cross-platform and streamed services for a variety of devices. These differences enable edge computing to support broader IoT applications with various requirements.

As part of this convergence, IoT applications (such as sensor-based services) will be delivered on-demand through a cloud environment [39]. This extends beyond the need to virtualize sensor data stores in a scalable fashion. It asks for virtualization of Internet-connected objects and their ability to become orchestrated into on-demand services (such as Sensing-as-a-Service).

Computing at the edge of the mobile network defines the IoT-enabled customer experiences and require a resilient and robust underlying network infrastructures to drive business success. IoT assets and devices are connected via mobile infrastructure, and cloud services are provided to IoT platforms to deliver real-time and context-based services. Edge computing is using the power of local computing and using different types of edge devices, to provide intelligent services. Data storage, computing and control can be separated and distributed among the connected edge devices (servers, micro servers, gateways, IoT nodes, etc.). Thus, edge computing advantages, such as, improved scalability, local processing, contextual computing and analytics. Interacting with the cloud (see Figure 3.25), edge computing provide scalable services for different type of IoT applications.

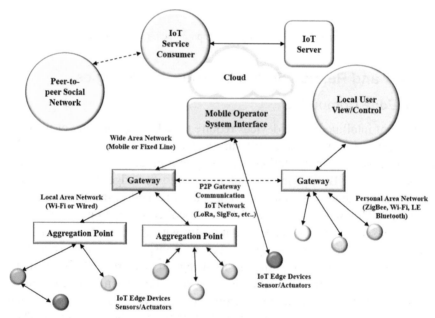

Figure 3.25   IoT centralised and distributed networks – gateway aggregation points.

Data transmission costs and the latency limitations of mobile connectivity pose challenges to many IoT applications that rely on cloud computing. Mobile edge computing will enable businesses to deliver real-time and context-based mobile moments to users of IoT solutions, while managing the cost base for mobile infrastructure. A number of challenges listed below have to be addressed when considering edge-computing implementation [40]:

- Cloud computing and IoT applications are closely connected and improve IoT experiences. IoT applications gain functionality through cloud services, which in turn open access to third-party expertise and up-to-date information.
- Mobile connectivity can create challenges for cloud-enabled IoT environments. Latency affects user experiences, so poor mobile connectivity can limit cloud-computing deployments in the IoT context.
- Mobile edge computing provides real-time network and context information, including location, while giving application developers and business leaders access to cloud computing capabilities and a cloud service environment that's closer to their actual users.
- Mobile edge computing is an important network infrastructure component for blockchain. The continuous replication of "blocks" via devices on this distributed data centre poses a tremendous technological challenge. Mobile edge computing reveals one opportunity to address this challenge.

For the future IoT applications it is expected that more of the network intelligence to reside closer to the source. This will push for the rise of Edge Cloud/Fog, Mobile Edge computing architectures, as most data will be too noisy or latency-sensitive or expensive to be transfer to the cloud. The edge computing technologies for IoT require to address issues such as unstable and intermittent data transmission via wireless and mobile links, efficient distribution and management of data storage and computing, edge computing interfacing with the cloud computing to provide scalable services, and finally mechanisms to secure the IoT applications. In this context, the research challenges in this area are:

- Open distributed edge computing architectures and implementations for IoT
- Modelling and performance analysis for edge computing in IoT
- Heterogenous wireless communication and networking in edge computing for IoT
- Resource allocation and energy efficiency in edge computing for IoT

- QoS and QoE provisioning in edge computing for IoT
- Trust, distributed end-to-end security and privacy issues in edge computing for IoT
- Federation and cross-platform, service supply in transparent computing for IoT

3.4.2 Networks and Communication

It is predicted that low-power short-range networks will dominate wireless IoT connectivity through 2025, far outnumbering connections using wide-area IoT networks [31], while 5G networks will deliver 1,000 to 5,000 times more capacity than 3G and 4G networks today. IoT technologies are extending the known business models and leading to the proliferation of different ones as companies push beyond the data, analytics and intelligence boundaries, while, everything will change significantly. IoT devices will be contributing to and strongly driving this development. Changes will first be embedded in given communication standards and networks and subsequently in the communication and network structures defined by these standards.

Network Technology

The development in cloud and mobile edge computing requires network strategies for fifth evolution of mobile the 5G, which represents clearly a convergence of network access technologies. The architecture of such network has to integrate the needs for IoT applications and to offer seamless integration and optimise the access to Cloud or mobile edge computing resources. IoT is estimated that will connect 30 billion devices. All these devices are connecting humans, things, information and content, which is changing the performance characteristics of the network. Low latency is becoming crucial (connected vehicles or industrial equipment must react in ms), there is a need to extend network coverage even in non-urban areas, a better indoor coverage is required, ultra-low power as many of the devices will be battery operated is needed and a much higher reliability and robustness is requested.

5G networks will deliver 1,000 to 5,000 times more capacity than 3G and 4G networks today and will be made up of cells that support peak rates of between 10 and 100 Gbps. They need to be ultra-low latency, meaning it will take data 1–10 milliseconds to get from one designated point to another, compared to 40–60 milliseconds today. Another goal is to separate communications infrastructure and allow mobile users to move seamlessly between 5G, 4G, and Wi-Fi, which will be fully integrated with the cellular network.

Applications making use of cloud computing, and those using edge computing will have to co-exist and will have to securely share data. The right balance needs to be found between cloud/mobile edge computing to optimize overall network traffic and optimize the latency. Facilitating optimal use of both mobile edge and cloud computing, while bringing the computing processing capabilities to the end user. Local gateways can be involved in this optimization to maximize utility, reliability, and privacy and minimize latency and energy expenditures of the entire networks.

Future networks have to address the interference between the different cells and radiations and develop new management models control roaming, while exploiting the co-existence of the different cells and radio access technologies. New management protocols controlling the user assignment to cells and technology will have to be deployed in the mobile core network for a better efficiency in accessing the network resource. Satellite communications need to be considered as a potential radio access technology, especially in remote areas. With the emerging of safety applications, minimizing the latency and the various protocol translation will benefit to the end to end latency. Densification of the mobile network strongly challenges the connection with the core network. Future networks should however implement cloud utilization mechanisms to maximize the efficiency in terms of latency, security, energy efficiency and accessibility.

In this context, there is a need for higher network flexibility combining Cloud technologies with Software Defined Networks (SDN) and Network Functions Virtualisation (NFV), that will enable network flexibility to integrate new applications and to configure network resources adequately (sharing computing resources, split data traffic, security rules, QoS parameters, mobility, etc.)

The evolution and pervasiveness of present communication technologies has the potential to grow to unprecedented levels in the near future by including the world of things into the developing IoT. Network users will be humans, machines, things and groups of them.

Communication Technology

The communication with the access edges of the IoT network shall be optimized cross domain with their implementation space and it shall be compatible with the correctness of the construction approach. Figure 3.26 IoT communication topologies across the architectural layers used for different configurations in IoT applications.

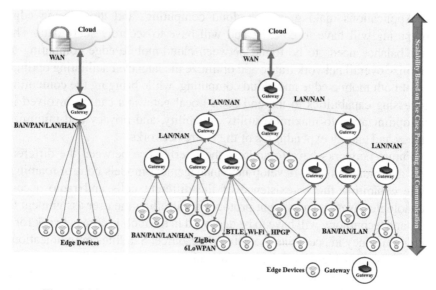

Figure 3.26 IoT Communication topologies across the architectural layers.

These trends require the extension of the spectrum in to the 10–100 GHz and unlicensed band and technologies like WiGig or 802.11ad that are mature enough for massive deployment, can be used for cell backhaul, point-to-point or point-to-multipoint communication. The use of advanced multi-/massive-MIMO technologies have the capability to address both coverage and bandwidth increase, while contributing to optimize the usage of the network resources adequately to real need.

Cisco expects by 2021 that 50% of all IP traffic will be Wi-Fi (30% will be carried by fixed networks and 20% via cellular networks). Parks Associates found that domestic smartphone users reported a 40% increase in their monthly Wi-Fi data consumption last year. Wi-Fi usage as presented in Figure 3.27 increased faster than mobile data usage, with two-thirds of smartphone users consuming more than 3 GB per month [13].

The IEEE's 802.11 (802.11k, 11r and 11v) standards focus on manageability, with features that address chaotic environments (such as transportation hubs) and steering Wi-Fi clients to less-congested, nearby access points (APs) automatically, depending on network conditions, as well as capabilities to better transition from AP to AP and network to network with very rapid handovers. The features of the latest 802.11 standards are shortly described below:

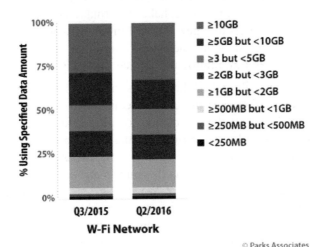

Figure 3.27   Wi-Fi Network – Mobile data consumed by network (2015–2016) [13].

- 802.11k – The 802.11k Assisted Roaming (AR) Roaming allows 11k capable clients to request a neighbor report containing information about known neighbor APs that are candidates for roaming. This feature enables the client to ask one AP about the other APs in the network – a neighbor list – and the client then makes use of that list in order to decide which AP to connect to, rather than sending probe signals.
- 802.11v – This feature enables client steering. A Wi-Fi client and an AP can both do some amount of measuring the signal environment and exchanging information to determine that "next best AP" for the client to connect to as it moves through the network. 802.11v enables an AP to warn and then force a client with a fading signal to disconnect, while providing information about other APs that the client can utilize that may be more lightly loaded or provide better signal strength. This prevents so-called "sticky" clients from being able to cling to a preferred AP that is providing a poor signal or is congested.
- 802.11r – The 802.11r Fast Transition (FT) Roaming uses a new concept for roaming. The initial handshake with the new Access Point (AP)

occurs before client roams to the target AP, called as Fast Transition (FT). In terms of authentication, once a device is authenticated on a Wi-Fi network, 802.11r provides some short-cuts on authentication among APs on the same network – so the network and device aren't basically trading the same "handshake" information back and forth every time a client connects to a new AP within the network. The number of exchanges is reduced, which means the roaming or reconnection time is also reduced, therefore improving the user experience.

The IoT applications will embed the devices in various forms of communication models that will coexist in heterogeneous environments. The models will range from device to device, device to cloud and device to gateway communications that will bring various requirements to the development of electronic components and systems for IoT applications. The first approach considers the case of devices that directly connect and communicate between each another (i.e. using Bluetooth, Z-Wave, ZigBee, etc.) not necessarily using an intermediary application server to establish direct device-to-device communications. The second approach considers that the IoT device connect (i.e. using wired Ethernet or Wi-Fi connections) directly to Internet cloud/fog service of various service providers to exchange data and control message traffic. The third approach, the IoT devices connect to an application layer gateway running an application software operating on the gateway device, providing the "bridge" between the device and the cloud service while providing security, data protocol translation and other functionalities. The gateway has a key role in the communications layer that integrates a collection of communication networks, which enable flow of information across the application domain system. No single technology will cover all aspects and needs of the various domains: the network requirements are largely driven by applications and use cases. The need of connecting anything from anywhere brings new scenarios that depend on latency, mission criticality, time-of-use pricing, peak data rates, security needs, battery life, and distance requirements, which allows various protocols and networks accessible for use. For some applications, the networking will allow connecting to ZigBee, wM-Bus, Power Line Communication (PLC), Wi-Fi, Z-Wave, LonWorks, while for other could be PLC, Ethernet Broadband, Ethernet IP, Wi-Fi/WiMax, 2G/3G/4G cellular or even satellite communication. Figure 3.28 illustrate the trade-off between range and data rates for the gateway and the different communication protocols.

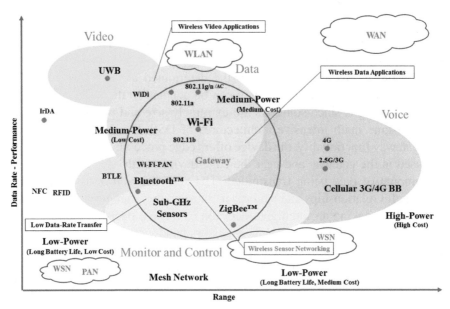

Figure 3.28 Gateway trade-off – data rate vs. range.

Gateway reference architecture allows for using the gateway reference designs in healthcare monitoring, environment parameters monitoring, indoor localization, wearables, by exploiting short-range/medium-range wireless connectivity. The individual rights to access secure information, secure data handling and privacy will be handled by security enabling components.

By delving into finer and more specific technical details, the targeted multi-functionality/multi-protocol reference gateways are complex embedded systems that require complex software running on high-end processor platforms, sometimes using real-time operating systems. The gateway reference designs manage edge heterogeneous devices and translate data across networks and into analytical cloud systems. They provide a customizable middleware development environment that provides security, connectivity, networking options, and device management to simplify the development, integration, and deployment of gateways for the IoT. In the residential environments, the gateways are emerging as integration platforms or an edge computing platform that enables to seamlessly interconnect various devices and other systems into a system of systems. The gateway enables users to securely aggregate, process, share, and filter data for analysis. The new gateways may include the integration of analytics into the gateway functionality to address specific problems in the embedded applications directly in the localities where analytics results may be promptly detected and trigger rapid reactions in a decentralized way.

The processing solutions to run the determined analytic algorithms in the most efficient and effective way can be adapted to the various gateway reference designs since solutions are extremely size and power constrained and real-time math intensive architectures of DSP. By including analytics into gateway functionality, the device offers edge processing, where analytics are used in the gateway and near the edge devices of the network to reduce the amount of data being transmitted. The analytics allows to discovering meaningful relationships and patterns in data and they can provide the ability to facilitate making an intelligent decision or make a decision based on the data that can provide a way to reduce network bandwidth by only transmitting the relevant information and not the entire data stream.

In addition, in order to increase portability, easy deployability, and easy extensibility in the heterogeneous and ever changing ecosystem of IoT components, virtualization and efficient cloud integration techniques can be used. The gateway design blend communications and computing technologies and integrates software-defined networking (SDN) and network functions virtualization (NFV) to consolidate network, cloud, and data centre functions onto standard, high-volume servers, switches, and storage. Several reference design implementations use general-purpose Java, Java-based OSGi, Eclipse Mihini, and Lua and support open standard IoT protocols, also for the efficient integration with heterogeneous cloud resources, including TR-069, OMA-DM/LWM2M, XMPP, CoAP, MQTT, while using hypervisor and virtualization technologies.

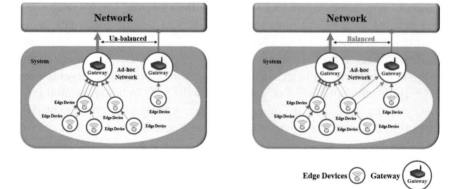

Figure 3.29 Network requirements – un-balance and balanced case.

The modular integrated connectivity creates a scalable mobile platform (modems for 2G/3G/4GLTE), enabling high-speed data and voice. and various onboard selected LoRa, Sigfox, On Ramp Wireless, NWave/ Weightless SIG, 802.11 Wi-Fi/Wi-Fi Aware, Bluetooth, ZigBee, 6LowPAN, Z-Wave, EnOcean, Thread, wMBus protocols using multiple ISM radio bands simultaneously (i.e. 169/433/868/902 MHz, 2.4 GHz, and 5 GHz), The connectivity modules are based on integrated ICs, reference designs, and feature-rich software stacks based on a flexible, modular concept that properly addresses various application domains.

The load of the network will be different with models using unbalanced load of the ad-hoc network from the core network point of view, while other using network-based solutions by balancing the topology from the core network point of view. In this case the identified network requirement to be supported are the calculation of the optimal ad-hoc network topology by using monitoring information, and notification of appropriate actions based on calculation results as presented in

The deployment of billions of devices requires network agnostic solutions that integrate mobile, narrow band IoT (NB-IoT), LPWA networks, (LoRA, Sigfox, Weightless, etc) as presented in Figure 3.30, and high speed wireless networks (Wi-Fi), particularly for applications spanning multiple jurisdictions.

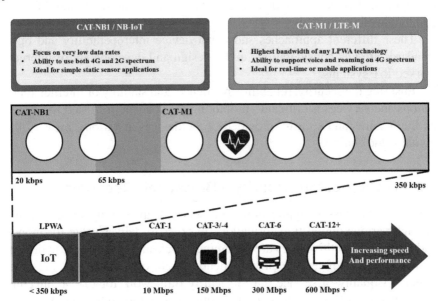

Figure 3.30 LPWA, NB-IoT and LTE-M for low data rates IoT applications.

LPWA networks have several features that make them particularly attractive for IoT devices and applications that require low mobility and low levels of data transfer:

- Low power consumption that enable devices to last up to 10 years on a single charge
- Optimised data transfer that supports small, intermittent blocks of data
- Low device unit cost
- Few base stations required to provide coverage
- Easy installation of the network
- Dedicated network authentication
- Optimised for low throughput, long or short distance
- Sufficient indoor penetration and coverage

These different types of networks are needed to address IoT product, services and techniques to improve the Grade of Service (GoS), Quality of Service and Quality of Experience (QoE) for the end users. Customization-based solutions, are addressing industrial IoT while moving to a managed wide-area communications system and, ecosystem collaboration.

Intelligent gateways will be needed at lower cost to simplify the infrastructure complexity for end consumers, enterprises, and industrial environments. Multi-functional, multi-protocol, processing gateways are likely to be deployed for IoT devices and combined with Internet protocols and different communication protocols.

These different approaches show that device interoperability and open standards are key considerations in the design and development of internetworked IoT systems.

Ensuring the security, reliability, resilience, and stability of Internet applications and services is critical to promoting the concept of trusted IoT based on the features and security provided of the devices at various levels of the digital value chain.

3.5 IoT Distributed Security – Blockchain Technology

IoT-based businesses, applications and services are scaling up and going through various digital transformations in order to deliver value for money and remain competitive, they are becoming increasingly vulnerable to disruption from denial-of-service attacks, identity theft, data tampering and other threats. A quality-of-service (QoS) security framework for IoT architecture is presented in [2], comprising of authentication, authorization, network and trust management components.

The IoT must embrace distributed technologies in order to both scale and provide end-to-end security, trust and accountability. In [1], swarm intelligence (SI), a subfield of artificial intelligence (AI), is presented as a source of inspiration for the design of new IoT security solutions. The use of SI makes it possible to add both cognitive and collective intelligence to IoT objects. Thus, IoT objects will strive to improve to a higher level of local intelligence in order to fulfil their function in a distributed manner, while the collective intelligence is centralised in order to solve problems that are more complex.

IoT objects are becoming more intelligent and more capable of making decisions both individually and collectively; they are also driven by collaboration, collective efforts and competition. Thus, the consensus problem, which is fundamental to decision-making in multi-agent systems, has received attention recently – and not least due to the newly emerging blockchain technology.

Blockchain technology addresses trust and security issues in an open and transparent manner, allowing the democratisation of trust. This is achieved by maintaining a record of every transaction made by every participant and having many participants verify each transaction, thus providing highly redundant verification and eliminating the need for centralised trust authorities. Consensus is thus achieved in a more effective way [3].

The concept of distributed consensus is at the core of blockchain technology [7]. Distributed consensus in the digital world means various nodes in a network coming to an agreement in a way that is very similar to a group of people, i.e., each member contributes their own opinion, and the group as a whole comes to a collective decision (except of course that the implementation in the digital process is a rather complex computer science problem). Nevertheless, the success of the concept lies precisely in the fact that both the collective decision and the process are recorded. Any time a transaction is challenged, for example, if an abnormal behaviour can be traced back to it, the recording is able to provide proof of the context in which the transaction had been performed.

Thus, the rationale behind blockchain technology keeping records of both the decision and the process is to ensure verifiability, so that all past and current transactions can be verified at any time in the future. This allows a high degree of accountability, without compromising the privacy of either the digital assets or the parties involved. Anonymity is therefore another important feature of blockchain technology.

Anonymity is difficult to achieve, even with a group of people seeking consensus, due to the very basic principles that consensus is built upon: members engage in dialogue and share information for the purpose of increasing the group's understanding of the issues, thus providing a rationale for each member's particular opinion, position and, ultimately, the group's collective decision. The same level of interaction is not required in the case of majority rule, and so people can more easily remain anonymous.

Although the concept of the blockchain is linked to Bitcoin, it is applicable to any digital asset exchanged online, where the asset is not money, but information. An example is the healthcare sector, where information is highly confidential. With the advance of technology, patients are located more and more often outside of hospitals and medical centres, and so their secure communication network involves more agents. Thus, the security of all transactions in the wider distributed network is essential.

Blockchain, the underlying technology of Bitcoin, is a 'chain of digital signatures', as described by Satoshi Nakamoto [4]. It enables parties connected to the same network to exchange information and other assets without the need for a third party to mediate the exchange. Blockchain technology entails a digital platform of distributed database technologies and protocols, which ensure that the transactions are irrefutable and that the information stored and shared is unalterable. The blockchain is copied to all parties in the network. Each transaction is verified and validated by a consensus of the parties. The transactions are arranged chronologically in blocks and are linked together so that each block embeds the history of all assets and decision processes since the first transaction. Therefore, it is almost impossible to generate a fraudulent transaction or the race against the other parties and the consensus that ultimately leads to the transaction being verified and validated. This provides a guaranteed protection against malicious interventions.

There is no universally agreed-upon definition of blockchains. They consist of various types, such as public or private. Blockchains also use different consensus mechanisms so that they can be adapted for use in a specific application field. Despite these differences, each implementation of blockchain technology invariably has a number of common features that distinguish it from other technologies. Blockchain technology is:

- Decentralized, meaning that there is no need for a central or other trusted authority to keep the data or supervise since each node has a copy of the entire database.

- Consensus-driven, meaning that new blocks are only added upon agreement that they are verified and validated.
- Anonymous, meaning that the identities of the nodes that participated in a current or past transaction are withheld but can be traced in case the transaction is challenged in the future.
- Unalterable, meaning that it is difficult, if not impossible, to change historical data simply because all nodes possess copies of the records.
- Time-stamped, meaning that the date and time a block is added to the blockchain is recorded.
- Programmable, meaning that transactions and other actions are executed only when certain conditions are met.

Although it is a promising alternative to established practices based on centralised control, blockchain technology still faces challenges in several areas, such as privacy, scalability, security, costs and integration.

Variations of the Bitcoin 'proof-of-work' consensus mechanism as well as novel designs have been proposed to deal with environments that depart from the idealistic assumption that the nodes in a network will always act in an honest and predictable way. 'Proof-of-stake', 'smart contracts', 'Byzantine consensus' and 'deposit-based consensus' are examples of such solutions [5]. Nevertheless, reaching a consensus in the context of dishonest and distrusting nodes, which is a challenge in the field of distributed computing, also remains difficult with blockchain technology.

Despite the challenges, it is relatively easy to achieve one goal: making people trust the blockchain technology. It is a matter of common sense to question the integrity of a service that works without any trusted central authority. One way to earn people's confidence is through transparency, and exposing the technology's internal workings. Another way to achieve this is to prove the legitimacy of the technology for applications other than Bitcoin. To demonstrate the former approach, this chapter briefly addresses two aspects of the internal workings of blockchain-namely, 'smart contracts' and block verification and validation. Regarding the latter means of promoting trust, insights into the use of blockchain in healthcare are provided.

3.5.1 Verification and Validation in Blockchain

One thing not easily understood about blockchain technology is how a newly created transaction is verified and validated using distributed technologies. In the same way that people need a favourable environment to reach a consensus in the real world, a specific set of conditions is also required in the digital, online world. Although it is not easy, blocks can be generated fraudulently; therefore, it is necessary to decide which blocks to trust. The debate process that leads to a decision needs better exposure. What are the algorithms used to reach a consensus so that a transaction can be verified and validated?

A transaction created to initiate a service is represented as a 'block' and contains, at a minimum, a unique identifier and the source, destination, type and amount of the asset being exchanged. The block thus generated is broadcast to the rest of the network. The nodes perform verification and validation activities, based on which the transaction is accepted or rejected. If accepted the 'block' is added to the blockchain. What verification and validation mean and how much information is necessary to accept a transaction vary from node to node and from service to service. Depending on the results from the nodes, a consensus algorithm is needed to produce a decision.

Verification activities mainly consist of checking digital signatures and the relationship between each virtual node and the actor behind it. Biometrics are a reliable form of verification, and embedding biometrics into the blockchain or linking the blockchain to a biometrics database has already been proposed.

Validation activities focus on the business logic and distinguish between public and private blockchains. While Bitcoin uses miners, private networks use individual validators, or trusted actors whose identity is known to the rest of the network. The incentive for transaction validators is not in the form of bitcoins but rather in being part of the network and benefiting from its information and services [8]. The more validators are in a network, the more decentralised and credible its decisions will be.

3.5.2 IoT Blockchain Application in Healthcare

Another way to earn people's confidence in the blockchain technology is to prove the legitimacy of the technology for applications other than Bitcoin. There are multiple applications of blockchain in the healthcare industry, with the potential to offer innovative solutions to challenges caused by the increasing volume of patient data managed by various stakeholders. This in turn increases the vulnerability to hacking and ransomware attacks, as those which recently targeted hospitals in several European countries and the US. As the purpose of such attacks is to deprive the medical institution of its own data by locking this data, this would be more difficult to achieve if the data is distributed and replicated across many nodes in a blockchain-based network.

Another application is remote healthcare, which is a new way of managing the care of patients with long-term conditions outside hospitals, either because there are no hospitals in the geographic area or in order to reduce costs. Innovative solutions are required in order to place the individual at the centre of healthcare. The individual owns his/her own health and can access a life-time supply of records, anytime, anywhere. When necessary, the records are shared, accessed and interpreted by various stakeholders in a transparent, efficient and accountable manner. The blockchain represents the individual's healthcare path through life.

Another aspect of blockchain technology that is not easily understood is the new trend of smart contracts, which seems to mean different things to different people. Smart contracts are a self-executing code on a blockchain that automatically implements the terms of an agreement between parties.

In the healthcare industry, with the growth of connected health devices performing various functions, an Internet of Medical Things (IoMT) has started to evolve. The interoperability of the exchanged data, together with data security, privacy and reliability, must be a top priority, and blockchain and smart contracts have the potential to offer good solutions.

A proof-of-concept, which demonstrated how the principles of decentralisation and blockchain technology can be applied in the healthcare industry, has recently been reported in the literature [6]. The concept gives patients a comprehensive, immutable log as well as easy access to their medical information across providers and treatment sites. Medical stakeholders (researchers, public health authorities, etc.) are incentivised to participate in the network as blockchain 'miners'. The concept of contracts is depicted in Figure 3.31, where three type of contracts are shown on the left-hand side: Registrar Contracts (RC), which deal with identity registration and authentication; Patient-Provider Relationship Contracts (PPR), which deal with authorisation and the relationship between the patient and provider and fine-grained access control to medical data and, finally, Summary Contracts (SC), which deal with aggregate data for patients and providers. The blockchain implements references and pointers, allowing navigation in the relationship graphs between contracts and network nodes, examples of which are shown on the right-hand side.

On top of blockchain technology, smart contracts can be the solution to the integration of IoT and Artificial Intelligence (AI). Smart contracts are in fact codes that can trigger, for instance, a rule-based reasoning when certain conditions are met. Rules and conditions are pre-defined, and the AI techniques and methods used are limited mostly to the creation of intelligent

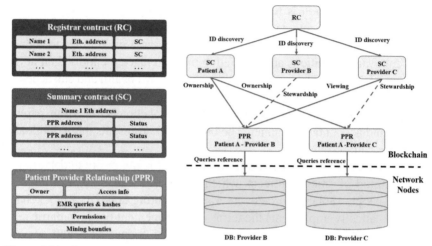

Figure 3.31   Smart contracts on the left-hand side and relationship graph between contracts and network nodes on the right-hand side [6].

representations of the nodes and existing medical records. However, it is envisaged that in the near future, more advanced AI techniques will be adopted so that nodes could learn to reason and act both independently and in groups in a more intelligent manner. In other words, these techniques will develop both local and collective intelligence. The traditional blockchain will become a cognitive blockchain, able to learn from past cases and adapt over time.

Blockchain technology augmented with both local and collective intelligence will make it possible for IoT objects to reason about household energy consumption, monitor and remotely collect patients' health status data, take actions through IoT devices and much more. Relevant use cases for the integration of these cutting-edge technologies are being reported in the literature, contributing to increasing levels of trust in them. The ultimate goal is to improve quality of life, whether by optimising household energy consumption, the use of self-driving cars, access to healthcare and so on.

3.6 IoT Platforms

IoT refers to an ecosystem in which applications and services are driven by data collected from devices that sense and interface with the physical world. Important IoT application domains span almost all major economic sectors: health, education, agriculture, transportation, manufacturing, electric grids, and many more.

IoT platforms enable companies to bring IoT solutions rapidly to the market by cutting development time and expenses for IoT systems.

An IoT Platform can be defined as an intelligent layer that connects the things to the network and abstract applications from the things with the goal to enable the development of services. The IoT platforms achieve a number of main objectives such as flexibility (being able to deploy things in different contexts), usability (being able to make the user experience easy) and productivity (enabling service creation in order to improve efficiency, but also enabling new service development). An IoT platform facilitates communication, data flow, device management, and the functionality of applications. The goal is to build IoT applications within an IoT platform framework. The IoT platform allows applications to connect machines, devices, applications, and people to data and control centres. The functionally of IoT platforms covers the digital value chain of an end-to-end IoT system, from sensors/actuators, hardware to connectivity, cloud and applications as illustrated in Figure 3.32. Different types of platforms have emerged [14].

IoT platforms' functionalities covers the digital value chain from sensors/actuators, hardware to connectivity, cloud and applications. Hardware connectivity platforms are used for connecting the edge devices and processing the data outside the datacentre (edge computing/fog computing), and program the devices to make decisions on the fly. The key benefits are security, interoperability, scalability and manageability by using advanced data management and analytics from sensor to datacentre. IoT software platforms

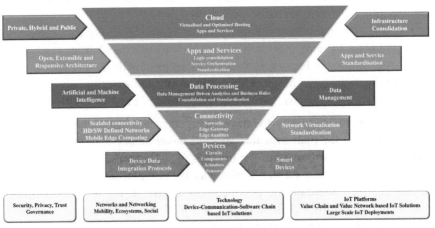

Figure 3.32 IoT Platforms covering the data value chain [14].

include the integration of heterogeneous sensors/actuators, various communication protocols abstract all those complexities and present developers with simple APIs to communicate with any sensor over any network. The IoT platforms also assist with data ingestion, storage, and analytics, so developers can focus on building applications and services, which is where the real value lies in IoT. Cloud based IoT platforms are offered by cloud providers to support developers to build IoT solutions on their clouds. Infrastructure as a Service (IaaS) providers and Platform as a Service (PaaS) providers have solutions for IoT developers covering different application areas. PaaS solutions, abstract the underlying network, compute, and storage infrastructure, have focus on mobile and big data functionality, while moving to abstract edge devices (sensors/actuators) and adding features for data ingestion/processing and analytics services [14]. The functions offered for the IoT consumer/business/industrial platforms are presented in Figure 3.33.

The IoT Platforms provide a framework for categorizing the technology capabilities that are necessary to deliver connected things, operations, assets, and the enterprises [14].

The four main blocks of capabilities presented in Figure 3.33 are:

- Connectivity that includes the hardware and software to network within the factory and the enterprise, standards for integrating machines, clouds, applications and the technology for managing devices, transferring data, and triggering events;
- Data analytics: including the use of a set of statistical and optimization tools to refine, monitor, and analyse structured and unstructured data for enabling different services;
- Cloud that integrate various types of cloud technologies across the enterprise to implement computing and storage capabilities (i.e. at the edge, within the factory, at the enterprise, or outside the firewall);
- Application area that integrates the tools for creating new mashup software applications that leverage the areas of the IoT platform.

Various types of IoT platforms have emerged in the last past years as they developed from the platforms that specific stakeholders or industrial sectors have promoted. A summary of the findings in [14] is presented below.

The IoT platforms on the market can be grouped into four categories: device centric communication/connectivity centric IoT platforms, industry centric IoT platforms and cloud centric IoT platforms. Non-commercial open source platforms are emerging and gaining in importance in several domains.

APPLICATION

- Integrated Development Environment: Java, HTML5
- IoT Data Model and Execution Engine
- Workflow and Business Logic Modeller
- Collaboration, Social
- Mobile
- Search
- Security - Authentication, Access Control, Configuration management, Cryptography, Logging, Compliance

CLOUD

- Private/Public/Hybrid
- IaaS - Compute, Storage, Network
- PaaS - Run Time, Queue, Traditional DB/DW ; Data Historian ; In-Memory Database ; Hadoop/Data Lake
- SaaS - Traditional Enterprise Applications, Next generation IoT Enabled Applications
- Security - Authentication, Access Control, Configuration Management, Antivirus/Spyware, Cryptography, Logging, Data Tagging, Compliance

DATA ANALYTICS

- Statistical Programming: R, SAS, SPSS
- Search, Text Mining, Data Exploration
- Analytics: Image/Video, Time Series, Geospatial, Predictive Modelling, Machine learning, etc.
- Statistical Process Control
- Optimizing and Simulation
- Metrics and KPIs
- Visualisation

CONNECTIVITY

- Network Infrastructure - Wired, Wi-Fi, and Cellular
- Standards - Serial/Proprietary > Ethernet/Open
- M2M/Data acquisition - Enabled, Gateways, APIs, Web Services, OPCUA, Modbus TCP/IP, MQTT, etc.
- Device Management
- Complex Event Processing
- Alarms, Condition Based Monitoring
- Data transport and Speed
- Security - Authentication, Access Control, Intrusion Detection/Prevention, Firewalls, Application Whitelisting, Antivirus/Spyware, Cryptography, Logging, Data Tagging, Compliance, etc.

Figure 3.33   Implementation elements in the main areas covered by IoT platforms [14].

The device centric IoT platforms are developed as hardware-specific software platforms supported by companies that commercialize IoT device components and have built a software backend that is referred to as an IoT platform. These backends are often reference implementations to ease the development of end-to-end IoT solutions, which are made available as starting points to other ecosystem partners.

The connectivity IoT platforms address the connectivity of connected IoT devices via communication networks. The starting points are often traditional M2M platforms for connectivity and device management, and then these are evolving into platforms that provide support for the management of the full IoT service life cycle. Connectivity based platforms primarily focus on providing out of the sandbox solutions for device/product manufacturers, which they can drop into their existing products to make them connected. New platforms provide analytics tailored for extracting the business insights about the performance of connected devices.

The cloud centric IoT platforms are offerings from larger cloud providers, which aim to extend their cloud business into the IoT. They offer different solutions with for example Infrastructure-as-a-service IaaS back ends that provide hosting space and processing power for applications and services. The back ends used to be optimized for other applications have been updated and integrated into IoT platforms offerings from large companies.

The industrial centric IoT platforms are the platforms designed to address the challenges of industrial IoT and integrates extensive features compared with the IoT consumer and business solutions (i.e. strong integrated IT and OT end-to-end security framework).

As the IoT platforms landscape is developing very fast with companies applying different strategies and business models such as sectorial approach that starts with the connectivity layer and is extending to expend to a platform features from the bottom-up.

Large companies use the top-down approach that they have a portfolio of software platform or cloud services and build extension to address the specific requirements for IoT applications. The development is starting from the analytics and cloud part and developing out the IoT platform features from the top-down. In the industrial sector, there are different strategies with companies developing their own industrial IoT platform or using and partnership approach y building alliances to offer the full industrial IoT platform suite. A different approach is developing or extending the IoT platforms offers through targeted acquisitions and/or strategic mergers. Another strategy used by several companies is to use the tactical/strategic investments throughout the IoT ecosystem developed around their IoT platforms and technologies.

Open source platforms are predominately emerging in consumer IoT space, such as the home automation sector or are outcomes of collaborative IoT research initiatives. The main driver is the cumbersome integration of an increasingly diverse set of end devices and protocols – making it costly for proprietary platform providers.

The IoT developments in the last few years have generated multiple architectures, standards and IoT platforms and created a highly fragmented IoT landscape creating technological silos and solutions that are not interoperable with other IoT platforms and applications. In order to overcome the fragmentation of vertically-oriented closed systems, architectures and application areas and move towards open systems and platforms that support multiple applications, for enhancing the architecture of IoT open platforms by adding a distributed topology and integrating new components for integrating evolving sensing, actuating, energy harvesting, networking and interface technologies.

The key technological shift is to provide tools and methods for implementing components and mechanisms in different architectural layers that operates across multiple IoT architectures, platforms and applications contexts and add functionalities for actuation and smart behaviour.

The developments of IoT platforms need to evolve the distributed architecture concept, the methods and tools for IoT open platforms, including software/hardware components, which provide connectivity and intelligence, actuation and control features while linking to modular and ad-hoc cloud and edge services. The IoT open platforms architecture need to allow data analytics and open APIs as well as semantic interoperability across use cases and the federation of heterogeneous IoT systems across the full technology stack. Cloud- and edge based storage and data analytics, and smart applications running on the cloud and at the edge on intelligent sensing/actuating devices (i.e. autonomous vehicles, autonomous devices, robotic things, etc.).

The new concepts need to integrate hardware- and software- level security capabilities to create redundancies to prevent intrusions and enable robust, secure, trusted IoT end-to-end solutions using blockchain technology to facilitate the implementation of decentralized open IoT platforms that assure secured and trusted data exchange as well as record keeping.

The blockchain can serve as the general ledger, keeping a trusted record of all the messages exchanged between smart devices in a decentralized IoT topology.

The IoT platforms need to advance the existing IoT open platforms to enable both integration and federation of existing IoT mechanisms, solutions, and platforms, thus leveraging the exploitation of existing IoT systems while ensuring compatibility with existing developments addressing object identity management, discovery services, virtualisation of objects, devices and infrastructures and trusted IoT approaches.

Acknowledgments

The IoT European Research Cluster – European Research Cluster on the Internet of Things (IERC) maintains its Strategic Research and Innovation Agenda (SRIA), taking into account its experiences and the results from the on-going exchange among European and international experts.

The present document builds on the 2010, 2011, 2012, 2013, 2014, 2015 and 2016 Strategic Research and Innovation Agendas.

The IoT European Research Cluster SRIA is part of a continuous IoT community dialogue supported by the EC DG Connect – Communications Networks, Content and Technology, E4 – Internet of Things Unit for the European and international IoT stakeholders. The result is a lively document that is updated every year with expert feedback from on-going and future projects financed by the EC. Many colleagues have assisted over the last few years with their views on the IoT Strategic Research and Innovation agenda document. Their contributions are gratefully acknowledged.

List of Contributors

Abdur Rahim Biswas, IT, CREATE-NET, WAZIUP
Alessandro Bassi, FR, Bassi Consulting, IoT-A, INTER-IoT
Alexander Gluhak, UK, Digital Catapult, UNIFY-IoT
Amados Daffe, SN/KE/US, Coders4Africa, WAZIUP
Antonio Skarmeta, ES, University of Murcia, IoT6
Arkady Zaslavsky, AU, CSIRO, bIoTope
Arne Broering, DE, Siemens, BIG-IoT
Bruno Almeida, PT, UNPARALLEL Innovation, FIESTA-IoT, ARMOUR, WAZIUP
Carlos E. Palau, ES, Universitat Politcnica de Valencia, INTER-IoT
Charalampos Doukas, IT, CREATE-NET, AGILE
Christoph Grimm, DE, University of Kaiserslautern, VICINITY
Claudio Pastrone, IT, ISMB, ebbits, ALMANAC
Congduc Pham, FR, Universite de Pau et des Pays de l'Adour, WAZIUP
Elias Tragos, IE, Insight Centre for Data Analytics, NUIG and FORTH-ICS, RERUM, FIESTA-IoT
Eneko Olivares, ES, Universitat Politcnica de Valencia, INTER-IoT
Fabrice Clari, FR, inno TSD, UNIFY-IoT
Franck Le Gall, FR, Easy Global Market, WISE IoT, FIESTA-IoT, FESTIVAL

Frank Boesenberg, DE, Silicon Saxony Management, UNIFY-IoT
François Carrez, UK, University of Surrey, FIESTA-IoT
Friedbert Berens, LU, FB Consulting S.à r.l, BUTLER
Gabriel Marão, BR, Perception, Brazilian IoT Forum
Gert Guri, IT, HIT, UNIFY-IoT
Gianmarco Baldini, IT, EC, JRC
Giovanni Di Orio, PT, UNINOVA, ProaSense, MANTIS
Harald Sundmaeker, DE, ATB GmbH, SmartAgriFood, CuteLoop
Henri Barthel, BE, GS1 Global
Ivana Podnar, HR, University of Zagreb, symbIoTe
JaeSeung Song, KR, Sejong University, WISE IoT
Jan Höller, SE, EAB
Jelena Mitic DE, Siemens, BIG-IoT
Jens-Matthias Bohli, DE, NEC
John Soldatos, GR, Athens Information Technology, FIESTA-IoT
José Amazonas, BR, Universidade de São Paulo, Brazilian IoT Forum
Jose-Antonio, Jimenez Holgado, ES, TID
Jun Li, CN, China Academy of Information and Communications
Technology, EU-China Expert Group
Kary Främling, FI, Aalto University, bIoTope
Klaus Moessner, UK, UNIS, IoT.est, iKaaS
Kostas Kalaboukas, GR, SingularLogic, EURIDICE
Latif Ladid, LU, UL, IPv6 Forum
Levent Gürgen, FR, CEA-Leti, FESTIVAL, ClouT
Luis Muñoz, ES, Universidad De Cantabria
Manfred Hauswirth, IE, Insight Centre for Data Analytics, NUIG, OpenIoT,
VITAL
Marco Carugi, IT, ITU-T, ZTE
Marilyn Arndt, FR, Orange
Markus Eisenhauer, DE, Fraunhofer-FIT, HYDRA, ebbits
Martin Bauer, DE, NEC, IoT-A
Martin Serrano, IE, Insight Centre for Data Analytics, NUIG, OpenIoT,
VITAL, FIESTA-IoT, BIG-IoT
Martino Maggio, IT, Engineering – Ingegneria Informatica Spa, FESTIVAL,
ClouT
Maurizio Spirito, IT, Istituto Superiore Mario Boella, ebbits, ALMANAC,
UNIFY-IoT
Maarten Botterman, NL, GNKS, SMART-ACTION
Ousmane Thiare, SN, Université Gaston Berger, WAZIUP

Payam Barnaghi, UK, UNIS, IoT.est
Philippe Cousin, FR, FR, Easy Global Market, WISE IoT, FIESTA-IoT, EU-China Expert Group
Philippe Moretto, FR, ENCADRE, UNIFY-IoT, ESPRESSO, Sat4m2m
Raffaele Giaffreda, IT, CNET, iCore
Roy Bahr, NO, SINTEF, UNIFY-IoT
Sébastien Ziegler, CH, Mandat International, IoT6
Sergio Gusmeroli, IT, Engineering, POLIMI, OSMOSE, BeInCPPS
Sergio Kofuji, BR, Universidade de São Paulo, Brazilian IoT Forum
Sergios Soursos, GR, Intracom SA Telecom Solutions, symbIoTe
Sophie Vallet Chevillard, FR, inno TSD, UNIFY-IoT
Srdjan Krco, RS, DunavNET, IoT-I, SOCIOTAL, TagItSmart
Steffen Lohmann, DE, Fraunhofer IAIS, Be-IoT
Sylvain Kubler, LU, University of Luxembourg, bIoTope
Takuro Yonezawa, JP, Keio University, ClouT
Toyokazu Akiyama, JP, Kyoto Sangyo University, FESTIVAL
Veronica Barchetti, IT, HIT, UNIFY-IoT
Veronica Gutierrez Polidura, ES, Universidad De Cantabria
Xiaohui Yu, CN, China Academy of Information and Communications Technology, EU-China Expert Group

Contributing Projects and Initiatives

SmartAgriFood, EAR-IT, ALMANAC, CITYPULSE, COSMOS, CLOUT, RERUM, SMARTIE, SMART-ACTION, SOCIOTAL, VITAL, BIG IoT, VICINITY, INTER-IoT, symbIoTe, TAGITSMART, bIoTope, AGILE, Be-IoT, UNIFY-IoT, ARMOUR, FIESTA, ACTIVAGE, AUTOPILOT, CREATE-IoT, IoF2020, MONICA, SYNCHRONICITY, U4IoT.

List of Abbreviations and Acronyms

Acronym	Meaning
3GPP	3rd Generation Partnership Project
API	Application Programming Interface
ARM	Architecture Reference Model
Bluetooth	Proprietary short range open wireless technology standard
BUTLER	EU FP7 research project uBiquitous, secUre inTernet of things with Location and contExt-awaReness

CAGR	Compound annual growth rate
DoS/DDOS	Denial of service attack
	Distributed denial of service attack
EC	European Commission
ESOs	European Standards Organisations
ESP	Energy Service Provider
ETSI	European Telecommunications Standards Institute
EU	European Union
FP7	Framework Programme 7
GS1	Global Standards Organization
IBM	International Business Machines Corporation
ICT	Information and Communication Technologies
iCore	EU research project
	Empowering IoT through cognitive technologies
IERC	European Research Cluster for the Internet of Things
IETF	Internet Engineering Task Force
IoB	Internet of Buildings
IoE	Internet of Energy
IoT	Internet of Things
IoT6	EU FP7 research project
	Universal integration of the Internet of Things through an IPv6-based service oriented architecture enabling heterogeneous components interoperability
IoT-A	Internet of Things Architecture
IoT-I	Internet of Things Initiative
IoV	Internet of Vehicles
IP	Internet Protocol
IPv6	Internet Protocol version 6
LTE	Long Term Evolution
M2M	Machine to Machine
MIT	Massachusetts Institute of Technology
OASIS	Organisation for the Advancement of Structured Information Standards
OpenIoT	EU FP7 research project
	Part of the Future Internet public private partnership

	Open source blueprint for large scale self-organizing cloud environments for IoT applications
PAN	Personal Area Network
PET	Privacy Enhancing Technologies
PPP	Public-private partnership
PV	Photo Voltaic
SENSEI	EU FP7 research project
	Integrating the physical with the digital world of the network of the future
SmartAgriFood	EU ICT FP7 research project
	Smart Food and Agribusiness: Future Internet for safe and healthy food from farm to fork
SmartSantander	EU ICT FP7 research project
	Future Internet research and experimentation
SRIA	Strategic Research and Innovation Agenda
TC	Technical Committee
W3C	World Wide Web Consortium
ZigBee	Low-cost, low-power wireless mesh network standard based on IEEE 802.15.4
Z-Wave	Wireless, RF-based communications technology protocol

References

[1] O. Vermesan and P. Friess (Eds.). Digitising the Industry Internet of Things Connecting the Physical, Digital and Virtual Worlds, ISBN: 978-87-93379-81-7, River Publishers, Gistrup, 2016.

[2] O. Vermesan and P. Friess (Eds.). Building the Hyperconnected Society – IoT Research and Innovation Value Chains, Ecosystems and Markets, ISBN: 978-87-93237-99-5, River Publishers, Gistrup, 2015.

[3] Outlier Ventures Research, Blockchain-Enabled Convergence – Understanding The Web 3.0 Economy, online at https://gallery.mailchimp.com/65ae955d98e06dbd6fc737bf7/files/Blockchain_Enabled_Convergence.01.pdf

[4] What is a blockchain? https://www2.deloitte.com/content/dam/Deloitte/ch/Documents/innovation/ch-en-innovation-deloitte-what-is-blockchain-2016.pdf

[5] R. Chan. Consensus Mechanisms used in Blockchain. https://www.linke
din.com/pulse/consensus-mechanisms-used-blockchain-ronald-chan

[6] A. Ekblaw et al. A Case Study for Blockchain in Healthcare: "MedRec" prototype for electronic health records and medical research data, 2016 https://www.healthit.gov/sites/default/files/5-56-onc_blockchainchallenge_mitwhitepaper.pdf

[7] M. Crosby et al. BlockChain Technology. Beyond Bitcoin. Berkeley, University of California, 2015 http://scet.berkeley.edu/wp-content/uploads/BlockchainPaper.pdf

[8] G. S. Samman. How Transactions Are Validated On A Distributed Ledger, 2016. https://www.linkedin.com/pulse/how-transactions-validated-distributed-ledger-george-samuel-samman

[9] ITU-T, Internet of Things Global Standards Initiative, http://www.itu.int/en/ITU-T/gsi/iot/Pages/default.aspx

[10] International Telecommunication Union – ITU-T Y.2060 – (06/2012) – Next Generation Networks – Frameworks and functional architecture models – Overview of the Internet of things

[11] O. Vermesan, P. Friess, P. Guillemin, H. Sundmaeker, et al., "Internet of Things Strategic Research and Innovation Agenda", Chapter 2 in Internet of Things – Converging Technologies for Smart Environments and Integrated Ecosystems, River Publishers, 2013, ISBN 978-87-92982-73-5

[12] Yole Développement, Technologies & Sensors for the Internet of Things, Businesses & Market Trends 2014–2024, 2014, online at http://www.yole.fr/iso_upload/Samples/Yole_IoT_June_2014_Sample.pdf

[13] Parks Associates, Monthly Wi-Fi usage increased by 40% in U.S. smartphone households, online at https://www.parksassociates.com/blog/article/pr-06192017

[14] A. Gluhak, O. Vermesan, R. Bahr, F. Clari, T. Macchia, M. T. Delgado, A. Hoeer, F. Boesenberg, M. Senigalliesi and V. Barchetti, "Report on IoT platform activities", 2016, online at http://www.internet-of-things-research.eu/pdf/D03_01_WP03_H2020_UNIFY-IoT_Final.pdf.

[15] McKinsey & Company, Automotive revolution - perspective towards 2030. How the convergence of disruptive technology-driven trends could transform the auto industry, 2016

[16] IoT Platforms Initiative, online at https:// www.iot-epi.eu/

[17] IoT European Large-Scale Pilots Programme, online at https://european-iot-pilots.eu/

[18] Où porterons-nous les objets connectés demain?, online at http://lamontr econnectee.net/les-montres-connectees/porterons-objets-connectes-dem ain/

[19] S. Moore, (2016, December 7) Gartner Survey Shows Wearable Devices Need to Be More Useful, online at http://www.gartner.com/newsroom/ id/3537117

[20] Digital Economy Collaboration Group (ODEC), online at http://archive. oii.ox.ac.uk/odec/

[21] D. Maidment, Advanced Architectures and Technologies for the Development of Wearable Devices, White paper, 2014, online at https://www.arm.com/files/pdf/Advanced-Architectures-and-Technolog ies-for-the-Development-of-Wearable.pdf Accenture. Are you ready to be an Insurer of Things?, online at https://www.accenture.com/_acnmedi a/Accenture/Conversion-Assets/DotCom/Documents/Global/PDF/Strat egy_7/Accenture-Strategy-Connected-Insurer-of-Things.pdf#zoom=50

[22] Connect building systems to the IoT, online at http://www.electronics-know-how.com/article/1985/connect-building-systems-to-the-iot

[23] S. Kejriwal and S. Mahajan, Smart buildings: How IoT technol-ogy aims to add value for real estate companies The Internet of Things in the CRE industry, Deloitte University Press, 2016, on-line at https://www2.deloitte.com/content/dam/Deloitte/nl/Documents/ real-estate/deloitte-nl-fsi-real-estate-smart-buildings-how-iot-technolog y-aims-to-add-value-for-real-estate-companies.pdf

[24] ORGALIME Position Paper, 2016, online at http://www.orgalime.org/ sites/default/files/position-papers/Orgalime%20Comments_EED_EPBD _Review%20Policy%20Options_4%20May%202016.pdf

[25] J. Hagerman, U.S. Department of Energy, Buildings-to-grid technical opportunities, 2014, https://energy.gov/sites/prod/files/2014/03/f14/B2G _Tech_Opps–Intro_and_Vision.pdf

[26] S. Ravens and M. Lawrence, Defining the Digital Future of Utilities – Grid Intelligence for the Energy Cloud in 2030, Navigant Research White Paper, 2017, online at https://www.navigantresearch.com/research /defining-the-digital-future-of-utilities

[27] Roland Berger Strategy Consultants, Autonomous Driving, 2014, online at https://www.rolandberger.com/publications/publication_pdf/roland_ berger_tab_autonomous_driving.pdf

[28] Roland Berger Strategy Consultants, Automotive Disruption Radar – Tracking disruption signals in the automotive industry, 2017, online at

https://www.rolandberger.com/publications/publication_pdf/roland_berg
er_disruption_radar.pdf

[29] The EU General Data Protection Regulation (GDPR) (Regulation (EU) 2016/679).

[30] RERUM, EU FP7 project, www.ict-rerum.eu

[31] Gartner Identifies the Top 10 Internet of Things Technologies for 2017 and 2018, online at http://www.gartner.com/newsroom/id/3221818

[32] A Look at Smart Clothing for 2015, online at http://www.wearable-technologies.com/2015/03/a-look-at-smartclothing-for-2015/

[33] Best Smart Clothing – A Look at Smart Fabrics 2016, online at http://www.appcessories.co.uk/best-smart-clothing-a-look-at-smart-fabrics/

[34] C. Brunkhorst, "Connected cars, autonomous driving, next generation manufacturing - Challenges for Trade Unions", Presentation at IndustriAll auto meeting Toronto Oct. 14th 2015, online at http://www.industriall-union.org/worlds-auto-unions-meet-in-toronto

[35] Market research group Canalys, online at http://www.canalys.com/

[36] Digital Agenda for Europe, European Commission, Digital Agenda 2010-2020 for Europe, online at http://ec.europa.eu/information_society/digital-agenda/index_en.htm

[37] O. Vermesan, P. Friess, G. Woysch, P. Guillemin, S. Gusmeroli, et al., "Europe's IoT Stategic Research Agenda 2012", Chapter 2 in The Internet of Things 2012 New Horizons, Halifax, UK, 2012, ISBN 978-0-9553707-9-3

[38] O. Vermesan, et al., "Internet of Energy – Connecting Energy Anywhere Anytime" in Advanced Microsystems for Automotive Applications 2011: Smart Systems for Electric, Safe and Networked Mobility, Springer, Berlin, 2011, ISBN 978-36-42213-80-9

[39] M. Yuriyama and T. Kushida, "Sensor-Cloud Infrastructure – Physical Sensor Management with Virtualized Sensors on Cloud Computing", NBiS 2010: 1–8

[40] Mobile Edge Computing Will Be Critical For Internet-Of-Things And Distributed Computing, online at http://blogs.forrester.com/dan_bieler/16-06-07-mobile_edge_computing_will_be_critical_for_internet_of_thing s_and_distributed_computing

4

Internet of Robotic Things – Converging Sensing/Actuating, Hyperconnectivity, Artificial Intelligence and IoT Platforms

Ovidiu Vermesan[1], Arne Bröring[2], Elias Tragos[3], Martin Serrano[3], Davide Bacciu[4], Stefano Chessa[4], Claudio Gallicchio[4], Alessio Micheli[4], Mauro Dragone[5], Alessandro Saffiotti[6], Pieter Simoens[7], Filippo Cavallo[8] and Roy Bahr[1]

[1]SINTEF, Norway
[2]SIEMENS AG, Germany
[3]National University of Ireland Galway, Ireland
[4]University of Pisa, Italy
[5]Heriot-Watt University, UK
[6]Örebro University, Sweden
[7]Ghent University – imec, Belgium
[8]Scuola Superiore Sant'Anna, Italy

Abstract

The Internet of Things (IoT) concept is evolving rapidly and influencing new developments in various application domains, such as the Internet of Mobile Things (IoMT), Autonomous Internet of Things (A-IoT), Autonomous System of Things (ASoT), Internet of Autonomous Things (IoAT), Internet of Things Clouds (IoT-C) and the Internet of Robotic Things (IoRT) etc. that are progressing/advancing by using IoT technology. The IoT influence represents new development and deployment challenges in different areas such as seamless platform integration, context based cognitive network integration, new mobile sensor/actuator network paradigms, things identification (addressing, naming in IoT) and dynamic things discoverability and many others. The IoRT represents new convergence challenges and their need to be addressed, in one side the programmability and the communication of multiple heterogeneous mobile/autonomous/robotic things for cooperating, their coordination, configuration, exchange of information, security, safety and protection. Developments in IoT heterogeneous parallel processing/communication and dynamic systems based on parallelism and concurrency require new ideas for integrating the intelligent "devices", collaborative robots (COBOTS), into IoT applications. Dynamic maintainability, self-healing, self-repair of resources, changing resource state, (re-) configuration and context based IoT systems for service implementation and integration with IoT network service composition are of paramount importance when new "cognitive devices" are becoming active participants in IoT applications. This chapter aims to be an overview of the IoRT concept, technologies, architectures and applications and to provide a comprehensive coverage of future challenges, developments and applications.

4.1 Internet of Robotic Things Concept

Artificial intelligence (AI), robotics, machine learning, and swarm technologies will provide the next phase of development of IoT applications.

Robotics systems traditionally provide the programmable dimension to machines designed to be involved in labour intensive and repetitive work, as well as a rich set of technologies to make these machines sense their environment and act upon it, while artificial intelligence and machine learning allow/empower these machines to function using decision making and learning algorithms instead of programming. The combination of these scientific disciplines opens the developments of autonomous programmable systems, combining robotics and machine learning for designing robotic systems to be autonomous.

Machine learning is part of an advanced state of intelligence using statistical pattern recognition, parametric/non-parametric algorithms, neural networks, recommender systems, swarm technologies etc. to perform autonomous tasks. In addition, the industrial IoT is a subset of the IoT, where edge devices, processing units and networks interact with their environments to generate data to improve processes [1]. It is in this area where autonomous functions and IoT can realistically allocate IoRT technology.

The use of communication-centred robots using wireless communication and connectivity with sensors and other network resources has been a growing and converging trend in robotics. A connected or "networked robot" is a robotic device connected to a communications network such as the Internet or LAN. The network could be wired or wireless, and based on any of a variety of protocols such as TCP, UDP, or 802.11. Many new applications are now being developed ranging from automation to exploration [64]. IEEE Society of Robotics and Automation's Technical Committee on Networked Robots [10] defines two subclasses of networked robots:

- Tele-operated robots, where human supervisors send commands and receive feedback via the network. Such systems support research, education, and public awareness by making valuable resources accessible to broad audiences.
- Autonomous robots, where robots and sensors exchange data via the network with minimum human intervention. In such systems, the sensor network extends the effective sensing range of the robots, allowing them to communicate with each other over long distances to coordinate their activity. The robots in turn can deploy, repair, and maintain the sensor network to increase its longevity, and utility.

A common challenge in the two subclasses of networked robots is to develop a science base that connect communication for controlling and enabling new capabilities, normally a robot is a closed system(s) with high capacities and where upgrades in functionality and operation (remote and/or local) requires expertise and usually long maintenance periods and where usually there is no open interfaces nor open communication channels and this is a way to guarantee security and control of efficiency.

Networked robots require wireless networks for sharing data among multiple robots, and to communicate with other, more powerful workstations used for computationally expensive and offline processing such as the creation of globally consistent maps of the robot's environment. This connectivity has strong implications for the sharing of tasks among robots, e.g. allowing tele-operation, as well as for human-robot interaction (HRI) and for on-the-fly reprogramming and adaptation of the robots on the network [16]. The evolution of these systems has now reached the consumer market, for instance, to support remote meetings and as tele-presence health-care tools. Cloud robotic systems have also emerged, to overcome the limitations of networked robotics through the provision of elastic resources from cloud infrastructure [9], and to exploit shared knowledge repositories over the Internet, making robots able to share information and learn from each other [34].

All these approaches pose several technical challenges related to network noise, reliability, congestion, fixed and variable time delay, stability, passivity, range and power limitations, deployment, coverage, safety, localization, sensor and actuation fusion, and user interface design. New capabilities arise frequently with the introduction of new hardware, software, and protocol standards.

The IoT technologies and applications are bringing fundamental changes in individuals' and society's view of how technology and business work in the world. Citizen centric IoT open environments require tackling new technological trends and challenges. In this context, the future developments where IoT infrastructure and services intersect with robotic and autonomous system technologies to deliver advanced functionality, along with novel applications, and new business models and investment opportunities, requires new IoT architectures, concepts and tools to be integrated into the open IoT platforms design and development.

The concept of IoRT goes beyond networked and collaborative/cloud robotics and integrates heterogenous intelligent devices into a distributed architecture of platforms operating both in the cloud and at the edge. IoRT addresses the many ways IoT today technologies and robotic "devices" convergence to provide advanced robotic capabilities, enabling aggregated IoT functionality along with novel applications, and by extension, new business, and investment opportunities [6] not only in industrial domains but in almost every sector where robotic assistance and IoT technology and applications can be imagined (home, city, buildings, infrastructures, health, etc.).

At the technology side, the proliferation of multi-radio access technology to connect intelligent devices at the edge has generated heterogeneous mobile networks that need complex configuration, management and maintenance to cope with the robotic things. Artificial intelligence (AI) techniques enable IoT robotic cognitive systems to be integrated with IoT applications almost seamlessly for creating optimized solutions and for particular applications. Cognitive IoT technologies allows embedding intelligence into systems and processes, allowing businesses to increase efficiency, find new business opportunities, and to anticipate risks and threats thus IoRT systems are better prepare to address the multiple requirements in the expected more IoT complex environment as it is depicted in Figure 4.1.

The combination of advanced sensing/actuating, communication, local and distributed processing, take the original vision for the IoT to a wholly different level, and one that opens completely new classes of opportunities for IoT and robotics solution providers, as well as users of their products. The concept enable baseline characteristics [1] that can be summarized as follow:

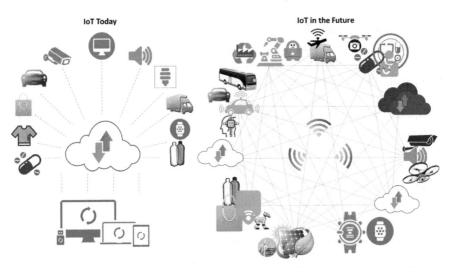

Figure 4.1 From a centralised cloud to distributed edge IoT platforms and applications.

- Define and describe the characteristics of robotics technologies that distinguish them as a separate, unique class of IoT objects, and one that differs considerably from the common understanding of IoT edge nodes as simple, passive devices.
- Reveal how the key features of robotics technology, namely movement, mobility, manipulation, intelligence and autonomy, are enhanced by the IoT paradigm, and how, in turn the IoT is augmented by robotic "objects" as "intelligent" edge devices.
- Illustrate how IoT and robotics technologies combine to provide for ambient sensing, ambient intelligence and ambient localization, which can be utilised by new classes of applications to deliver value.

IoT, cognitive computing and artificial intelligence technologies integration is part of the new developments foreseen for IoT applications in various smart environments.

4.2 Emerging IoRT Technologies

The definition of Internet of Things used in [3] states that IoT is "A dynamic global network infrastructure with self-configuring capabilities based on standard and interoperable communication protocols where physical and virtual "things" have identities, physical attributes, and virtual personalities and use intelligent interfaces, and are seamlessly integrated into the information network". The "things" are heterogeneous, have different levels of complexity, sensing/actuating, communication, processing, intelligence, mobility and are integrated into different platforms. The "robotic" things are a class of complex, intelligent, autonomous "things" that combine methods from robotics and from artificial intelligence [82] and are integrated to edge computing and cloud IoT based platforms. IoRT combines the features of a dynamic global network infrastructure with self-configuring capabilities with the autonomous, self-learning behaviour of connected robotic things creating a system of systems that learn itself using path- and motion-planning and motion control to create services and provide solutions to specific tasks. In this context, the IoT architecture integrates the autonomous system architecture based on six main characteristics:

- Sensing is a common characteristic of the IoT and Robotic systems and this is considered as the main characteristic to enable the interaction of devices "Things" with other IoT devices and people, most of the times only in the way device to human, from here the term "sensing", thus empowering people to be part of the ecosystem in the context of their IoT concept or service paradigm. This feature has been extensible investigated and "Sensing-as-a-Service" has been implemented among different solutions in IoT market.

- Actuating based on a holistic approach is the characteristics to enable devices "things" to action over physical and/or virtual activities, a feature or function that is well known in the IoT verticals but that is not currently available in the IoT open market. Actuating needs to look for a trusted, protected and secured development, deployment and operation of open multi-vendor IoT applications services. Actuating should be enabled on novel deployments as result of research efforts enabling "Actuation-as-a-Service" as a new paradigm for IoT enabling usability that ensure end user acceptance and engagement for controlled IoT devices.

- Control is an organised sequence of operations (mainly application layer) where functions and services are defined by a "loop" or a sequence of "loops" a.k.a. "Control Loops". The interfaces have to be defined to provide access to sensing information as well as to provide access to required control mechanisms and the comprehensive security concepts of the architecture have to be reflected in the interface definitions to enable the required sequencing mechanisms. The Control loop can be mapped virtually to anything, from applications to services in the cloud to networks devices in the networks infrastructure, if this last is possible then it is not difficult to believe the Internet of Things can be virtualize and represented by autonomic principles.

- Planning is an offered capability to orchestration-organize logic that coordinates the internal platform components for satisfying service requests and assuring that agreed quality levels are met throughout services life-cycle in the IoT application. The orchestration logic should align service requests with available resources, information handling and knowledge entities, and their platform-specific representation. Based on logic, planning relies on an automated workflow engine to instantiate the required functionality on a per service request basis. The orchestration logic will also maintain user-defined representations of information and resources to facilitate the process of service definition.

- Perception is known as the interdisciplinary approach in robotics where combining sensor information and knowledge modelling, robots aim to establish a robot-human interaction, by using human-interaction design, software engineering, service-based, cloud-based and data analytics architectures, multi-agent systems, machine sensor systems and sometimes artificial intelligence. Using perception robots become aware of the environment(s) enabling in this way a more particular activity for individual humans.

- Cognition, using this characteristic the device (robot) is intelligent in the sense that it has embedded monitoring (and sensing) capabilities and at the same time can get sensor data from other sources, which are fused for the "acting" purpose of the device. A second 'intelligent' part is that the device can leverage local and distributed "intelligence". In other words, it can analyse the data from the events it monitors (which means a presence of edge computing or fog computing in many circumstance) and has access to (analysed) data. Finally, both prior components serve the third one which consists of (autonomously) determining what action to take and when, whereby an action can be the control or manipulation of a physical object in the physical world and, if its purpose is to do so and it has been designed to be able to, the device or robot can also move in that physical world. In this stage 'notifying' or 'alerting', based upon the analysis of 'physical event' can be included.

The IoRT technologies that enable the development, implementation and deployment of IoRT applications are briefly described in the following subsections.

4.2.1 Sensors and Actuators

The two baseline technologies in IoT and robotics that are well defined and identified are sensors devices and actuators, both are always crucial components for implemented IoRT systems both with well-defined interfaces (e.g. for Identification or a Reaction) and for offering these functionalities to the IoRT platform via interaction components. Different from the IoT Sensors and Actuators compose the useful functionality in and out of the IoRT building blocks. Robotic Interaction Services (RoIS) defines also the use of external of the building block and abstracts the hardware in the service robot and the Human-Robot interaction (HRI) functions provided by the robot. Calling each of the HRI functions provided by a robotic system such as a service robot or an intelligent sensing system a "functional implementation", a robotic system can be expressed as a set of one or more functional sensor and actuator services implementations. These functional implementations (e.g. face recognition, wheel control) are usually provided in a form that is dependent on robot hardware such as sensors and actuators, examples of these sensors and actuators services are Radar, Lidar, Camera, Microphones, etc. HRI components (e.g. person detection, person identification) are logical functional elements, realized through physical units such as sensors placed on the robot and/or in the environment. The interesting part of this standard is that it allows to build applications that can be deployed on both gateways and devices, yet it is mainly focusing on HRI scenarios.

Robotic things inherit the potential for varied and complex sensing and actuation from the long tradition of robotics. From the sensing side, robotic science and technology provides methods and algorithms to use both simple and sophisticated sensors, including inertial sensors (accelerometer, compass, gyro), ranging sensors (sonar, radar, LIDAR – Light Detection and Ranging), 3D sensors (3D laser or RGBD camera), as well more common sensors like cameras, microphones and force sensors [79]. Mobile robots or multiple robots can collect sensor data from multiple pose and/or at multiple times, and techniques exist to combine these data in a coherent picture of the environment and of its evolution in time [80]. From the actuation side, the ability to modify the physical environment is arguably the most unique aspect of robotic things. Actuation can take a wide range of forms, from to operation of simple devices like an automatic door to the transportation of goods and people and to the manipulation of objects. An impressive range of techniques for actuation have been developed in the robotics field, including techniques for autonomous planning and execution of actions by single or multiple robots [81].

The IoRT applications require low-cost solid state semiconductor (CMOS) imaging sensors based on active illumination (laser based) that are robust in different environmental conditions such as sunlight, darkness, rain, fog, dust, etc. The sensors need to provide both road surface scanning (horizontal projection) and object detection (vertical projection) with high resolution and accuracy.

Current sensors mainly provide 2D sensing information and the sensors fusion (=environment model) is focused on 2D representation. Future IoRT functions require additional height information, 3D mapping and sensors/actuators fusion. The robotic things require a 3D environment model based on or adapted to existing/new sensor technologies to allow a highly accurate and reliable scene interpretation and collaboration with other robotic things, by finding the optimized representation of 3D environmental information as trade-off between resource demand and optimized performance.

The 360° vision in complex autonomous robotic things/vehicles is assure by LIDAR systems that provides the all-around view by using a rotating, scanning mirror. The LIDAR system provides accurate 3D information on the surrounding environment in order to enable the very fast decision-making needed for self-driving autonomous robotic thing, which is processed and used for object identification, motion vector determination, collision prediction, obstacle avoidance strategies.

In the case of close-in control, the LIDAR systems are not effective and the autonomous robotic things/vehicles need to equipped with radars. Operating frequency for the radar is usually in the range of 76–81 GHz, which is allocated for this use, has RF propagation characteristics, and provides the required resolution. Other advantages of the 76–81 GHz frequency range (79 GHz band) are that the radar devices are small, while the risk of mutual interference is reduced due the smaller emission power required. Radar scanning is a promising technology for collision avoidance, especially when the environment is obscured with smoke, dust, or other weather conditions.

4.2.2 Communication Technologies

The communication architecture of IoRT needs new approaches enabling shared real-time computation and the exchange of data streams (necessary for 3D-awareness and vision systems) combined with internal communication, and edge computing to enable the virtualization of functions on the existing computing engines, while enabling the ease of use of such infrastructures in many domains. The communication infrastructure and the IoRT external communication need to be able to perform time critical communication to ensure collision prevention becomes possible, thus heavily reducing accidents and collisions.

IoRT uses typically networking technologies for local robots operation and white spectrum frequencies assigned for remote operation. IoT uses machine to machine communication and implement on standards like 4G, Wi-Fi, Bluetooth, and emergent ones like LoRa and SIGFOX, Open challenges in IoRT is achieving interoperability and establishing services at this level which is much more challenging and requires semantic knowledge from different domains and the ability to discover and classify services of things in general. This is difficult to achieve mainly because the conditions in IoRT changes rapidly and is dependent on applications, locations and use cases.

Communication protocols are the backbone of IoRT systems and enable network connectivity and integration to applications. Different communication protocols as presented in Figure 4.2 are used by the edge devices and robotic things to exchange data over the network by defining the data exchange formats, data encoding, addressing schemes for devices and routing of packets from source to destination. The protocols used are 802.11 – Wi-Fi which includes different Wireless Local Area Network (WLAN) communication standards (i.e. 802.11a that operates in the 5 GHz band, 802.11b and 802.11g operate in the 2.4 GHz band, 802.11n operates in the 2.4/5 GHz bands, 802.11ac operates in the 5 GHz band and 802.11ad operates in the 60 GHz band). The standards provide data rates from 1 Mb/s to 6.75 Gb/s and communication range in the order of 20 m (indoor) to 100 m (outdoor).

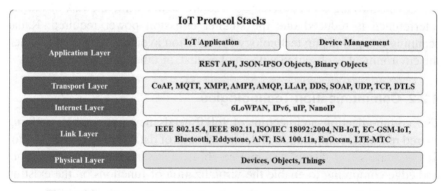

Figure 4.2   Communication protocols used by different IoRT applications.

The 802.15.4 – LR-WPAN IEEE 802.15.4 is a set of Low-Rate Wireless Personal Area Networks (LR-WPAN) standards based on the specifications for high level communications protocols such as ZigBee. LR-WPAN standards provide data rates from 40 Kb/s to 250 Kb/s. The standards provide low-cost and low-speed communication to power constrained devices and operates at 868/915 MHz and 2.4 GHz frequencies at low and high data rates, respectively.

The 2G/3G/4G and future 5G – mobile communication are different generations of mobile communication standards including second generation (2G including GSM and CDMA), third generation (3G-including UMTS, CDMA2000) and fourth generation (4G-including LTE).

IoT devices based on these standards can communicate over mobile networks with data rates ranging from 9.6 Kb/s (2G) to 100 Mb/s (4G).

The Narrowband IoT (NB-IoT) low power wide areas (LPWA) technology for IoT applications, use the existing 4G/LTE network and is based on 3GPP specifications [86]. The NB-IoT and LTE coexistence, the re-use of the LTE physical layer and higher protocol layers benefits the technology implementation. NB-IoT has been designed for extended range, and the uplink capacity can be improved in bad coverage areas. NB-IoT devices support three different operation modes [86]:

- Stand-alone operation: Utilizing one or more GSM carriers (bandwidth of 200 kHz replacements).
- Guard band operation: Utilizing the unused resource blocks within a LTE carriers' guard-band (frequency bands to prevent interference).
- In-band operation: Utilizing resource blocks within a normal LTE carrier.

For a wide range of applications, ten years battery lifetime and low cost devices will be available, and support a huge numbers of low-throughput things.

802.15.1 – Bluetooth is based on the IEEE 802.15.1 standard and offer a low power, low cost wireless communication technology for data transmission between mobile devices over a short range (8–10 m used in personal area network (PAN) communication. Bluetooth operates in 2.4 GHz band with data rate ranging from 1 Mb/s to 24 Mb/s. The ultra-low power, low cost version is called Bluetooth Low Energy (BLE which was merged with Bluetooth standard v4.0).

LoRaWAN R1.0 – LoRa is a long-range communication protocol that defines the Low Power Wide Area Networks (LPWAN) standard to enable IoT with data rates ranging from 0.3 kb/s to 50 kb/s. LoRa operates in 868 and 900 MHz ISM bands. LoRa communicates between the connected nodes within 30kms range, in unobstructed environments. The basis is the LoRa modulation, a wireless modulation for long-range radio, low power, low data rate applications, based on a chirp spread spectrum (CSS) technology. According to the LoRa Alliance [85], LoRa can demodulates signals 19.5 dB below the noise floor, while most frequency shift keying (FSK) systems need a signal power of 8–10 dB above the noise floor. Switching between LoRa CSS and FSK modulation are also facilitated. LoRaWAN is the network protocol optimized for battery-powered end-nodes. Battery life for the attached node is normally very long, up to 10 years.

The network server hosts the system intelligence and complexity (e.g., duplicate packets elimination, acknowledgement scheduling, data rate adapting). All connections are bidirectional, support multicast operation, and forms a star of stars topology. To serve different applications, the end-nodes are classified in three different classes, which trade off communication latency versus power consumption. Class A is the most energy efficient, and is implemented in all end-nodes. Class B and C are optional and must be class A compatible. A spreading factor (SF) is used to increase the network capacity. Higher SF gives longer communication range, but also imply decreased data rate and increased energy consumption. For frequent data sampling, LoRa systems use an SF as small as possible to limit the airtime, which requires end-nodes located closer to the gateways.

4.2.3 Processing and Sensors/Actuators Data Fusion

Connected robotic things can share their sensor data, fuse them, and reason collectively about them. The mobility and autonomy capabilities of robotic brings the problem of sensor fusion in IoT platforms to an entirely new level of complexity, and adds entirely new possibilities. Complexity is increased because of the great amount and variety of sensor data that robotic things can provide, and because the location of the sensing devices is not fixed and often is not know with certainty. New possibilities are enabled because of the ability of robotic things to autonomously move to specific locations to collect specific sensory input, based on the analysis of the currently available data and of the modelling and reasoning goals. The field of robotics has developed a wide array of technologies for multi-robot sensor fusion [65–67], as well as for active and goal-directed perception [68, 69]. These techniques would enable IoRT systems to dynamically and proactively collect wide ranges of data from the physical environment, and to interpret them in semantically meaningful ways.

4.2.4 Environments, Objects, Things Modelling and Dynamic Mapping

Robotic things need to maintain an internal model of their physical environment and of their own position within it. The model must be continuously updated to reflect the dynamicity of the environment. The problem of creating and maintaining this model while the position of the robots are changing is known as SLAM, for "simultaneous localization and map building", and it has been an active area of research in robotics for the past 20 years [70]. Techniques for metric 2D SLAM are now mature, and the field of robotics is now focusing on extending these techniques to build 3D maps [71], temporal dynamic maps [72], and semantic maps [73]. The latter are of special interest to IoRT systems, since they enrich purely metric information with semantic information about the objects and location in the environment, including their functionalities, affordances and relations.

4.2.5 Virtual and Augmented Reality

Robot-assisted surgery systems are applications that are integrating virtual reality (VR) and augmented reality (AR) technology in the operating room. Live and virtual imaging featured on robot-assisted user interfaces assist surgeon's manipulation of robotic instruments and represent an open platform for the addition of VR and AR capabilities. Live surgical imaging is used to enhance on robot-assisted surgery systems through image injection or the superimposition of location-specific objects. The application of VR/AR technology in robot-assisted surgery is motion tracking of robotic instruments within an interactive model of patient anatomy displayed on a console screen.

The techniques and technology can be extended to IoRT applications with fleets of robots using VR/AR for learning, navigation and supporting functions.

Augmented reality as technology enhances the real world by superimposing computer-generated information on top of it, augmented reality provides a medium in which digital information is overlaid on the physical world that is in both spatial and temporal registration with the physical world and that is interactive in real time [17].

The augmented reality tools allow cognitive robotics modelers to construct, at real-time, complex planning scenarios for robots, eliminating the need to model the dynamics of both the robot and the real environment as it would be required by whole simulation environments. Such frameworks build a world model representation that serves as ground truth for training and validating algorithms for vision, motion planning and control. The AR-based framework is applied to evaluate the capability of the robot to plan safe paths to goal locations in real outdoor scenarios, while the planning scene dynamically changes, being augmented by virtual objects [18].

4.2.6 Voice Recognition, Voice Control

Today, the conversational interfaces are focused on chatbots and microphone-enabled devices. The development of IoRT applications and the digital mesh encompasses an expanding set of endpoints with which humans and robotic things interact. As the IoRT mesh evolves, cooperative interaction between robotic things emerge, creating the framework for new continuous and ambient digital experience where robotic things and humans are collaborating.

The fleets of robots used in IoRT applications such as tour guiding, elder care, rehabilitation, search and rescue, surveillance, education, general assistance in everyday situations, assistants in factories, offices and homes require new and more intuitive ways for interactions with people and other robots using simple easy-to-use interfaces for human-robot interaction (HRI). The multimodality of these interfaces that address motion detection, sound localization, people tracking, user (or other person/robot) localization, and the fusion of these modalities is an important development for IoRT applications.

In this context, voice recognition and voice control requires robust methods for eliminating the noise by using information on the robot's own motions and postures, because a type of motion and gesture produces almost the same pattern of noise every time. The quality of the microphone is important for automatic speech recognition in order to reduce the pickup of ambient noise. The voice recognition control system for robots can robustly recognize voice by adults and children in noisy environments, where voice is captured using wireless microphones. To suppress interference and noise and to attenuate reverberation, the implementation uses a multi-channel system consisting of an outlier-robust generalized side-lobe canceller technique and a feature-space noise suppression criteria [19].

4.2.7 Orchestration

Smart behaviour and cooperation among sensing and actuating robotic things are not yet considered in the domains usually addressed with orchestration and dynamic composition of web-services in IoT platforms. An overview of middleware for prototyping of smart object environments was reported in [58]. The authors conclude that existing efforts are limited in the management of a huge number of cooperative SOs and that a cognitive-autonomic management is needed (typically agent-based) to fulfil IoT expectations regarding context-awareness and user-tailored content management by means of interoperability, abstraction, collective intelligence, dynamisms and experience-based learning. In addition, cloud and edge computing capabilities should complement the multi-agent management for data integration and fusion and novel software engineering methodologies need to be defined.

In general, existing IoT orchestration mechanisms have been designed to satisfy the requirements of sensing and information services – not those of physical robotic things sharing information and acting in the physical environment. Furthermore, these approaches cannot be directly mapped to embedded networks and industrial control applications, because of the hard boundary conditions, such as limited resources and real-time requirements [45]. Fortunately, robotic R&D has produced some prominent approaches to self-configuration of robotic networked robotic systems. Most noticeably, both the ASyMTRe system [40], and the system by Lundh et al. [41] consider a set of robots and devices, with a set of corresponding software modules, and define automatic ways to deploy and connect these modules in a "configuration" that achieves a given goal. These frameworks leverage concepts of classical planning, together with novel methods to reason about configurations for interconnecting modules. The approach by Lundh et al is more general, in that it considers highly heterogeneous devices, including simple wireless sensor network (WSN) nodes and smart objects. An extension of this approach, based on constraint-based planning [42], was developed in the FP7 projects RUBICON [43] and RobotEra [44]. The approach leverages an online planning and execution framework that incorporates explicit temporal reasoning, and which is thus able to take into account multiple types of knowledge and constraints characteristic of highly heterogeneous systems of robotic devices operating in open and dynamic environments.

4.2.8 Decentralised Cloud

One form of orchestration is computational harvesting, i.e. offloading of computational workload using decentralised cloud solutions. This can operate in two ways. First, from a resource-constrained device to an edge cloud. There is challenging energy-performance trade-off between on-board computation and the increased communication cost, while considering network latency [48]. This approach has been mainly studied in the context of offloading video processing workloads from smartphones and smart glasses [49]. AIOLOS is a middleware supporting dynamic offloading [50, 51], recently extended with a Thing Abstraction Layer, which advertises robots and IoT devices as OSGi-services that can be used in modular services [52].

Computational offloading has also found its way for robotics workloads. In the context of the H2020 MARIO (www.mario-project.eu) and H2020 RAPP (rapp-project.eu) projects, a framework was developed [59] where developers can create robotic applications, consisting of one Dynamic Agent (running on the robot) and one or more Cloud Agents. Cloud Agents must be delivered as a Docker container. The Dynamic Agents are developed in ROS, and need to implement a HOP web server to communicate with the Cloud Agents. Overall, the concept is mainly focused on offloading scenarios. For example, there is no support for public Cloud Agents: there is a one-to-one connection between a Cloud Agent and a Dynamic Agent. Targeted use cases are e.g. offloading of computationally intensive parts like SLAM. Similar work was done in the context of the European projects RoboEarth and follow-up RoboHow. All these frameworks are mainly oriented to allow the development of cloud-robot distributed applications and provide no integration or functionality for integration in the IoT [60].

Secondly, self-orchestration on edge clouds is related to the opposite direction, i.e. to shift (computational or storage) workloads from the centralized cloud closer to the endpoints (often the sources of data). This allows to reduce latency of control loops, or to mitigate the ingress bandwidth towards centralized servers, as recently specified by the Industrial Internet Consortium (IIC) for 3-tiers (edge, gateway, cloud) IoT architectures. Noticeable examples of such an approach include SAP Leonardo [53], GE Digital's Predix Machine [54]. IBM Watson IoT [55], and GreenGrass [56] by Amazon Web Services (AWS).

4.2.9 Adaptation

Current IoT platforms do not provide sufficient support for adaptability. Rather, adaptation must be addressed for each application, and usually relies on pre-programmed, static and brittle domain knowledge. This is further exacerbated in applications that need to smoothly adapt to hard-to-predict and evolving human activity, which is particularly the case for IoRT applications. Even with adaptation logic built-in the application, the only feasible approach is the applications leveraging on contextual knowledge and experience that is provided by the platform on which the application is deployed.

The need for adaptation is even more pronounced in an IoRT platform:

- Compared to sensor-based smart objects, the **number of contexts in which smart robotic things operate is a multiple**. A large share of robots is mobile and thus enters and leaves different operational contexts. These contexts may be demarcated by the communication range of sensors, by operational constraints (e.g. leaving a Wi-Fi access point, making some services inaccessible when connected to 4G). Also, non-mobile robots need to be flexibly reconfigured in terms of software and communication with other entities, e.g. in agile Industry 4.0 manufacturing. Future robotic things will be flexible in their actuation capabilities (i.e. not limited to a single pre-programmed functionality).
- While the co-habitation of multiple applications building on the same sensor data is conceptually straightforward (could be seen as the analogue to parallel reading operations of data in a OS), this claim is not sustainable in actuation (which could be somewhat seen as "write" operations). We see three different types of situations that may arise between actors in the IoRT: competitive (non-shareable, requires locking or reservation), cooperative (robots doing two tasks at the same time instead of executing them sequentially) and adversarial (two applications require opposite end-effects of the actuators).
- IoRT applications will often be deployed in large-scale environments which are open-ended in several dimensions: human expectations and preferences, tasks to be executed, number and type of (non-connected) objects that may appear in physical space. As argued above, adaptation in today's IoT (even when augmented with single-purpose actuators like smart automation) is a tedious procedure for which only limited platform support exists, but it must only be done once. In the IoRT, a more continuous adaptation is needed, because robots operate in open-ended, dynamic environments and are versatile actuators.
- Robotic devices are required to maintain a certain degree of autonomy. They should be given relatively high-level instructions ("Go to place X and deliver object Y i.e. they are not ideally suited for a more centralized orchestration approach to adaptation. These mandates a distributed setting with choreography between the different actors in the IoRT.

Considering all above elements, the IoRT objectives related to adaptation are truly novel. First, application developers must be provided with powerful tools to access *contextual learning* services that can provide up-to-date information and historic experience on the operational environment. Second, the platform must allow applications to self-configure in the distributed setting introduced above, i.e. by taking the responsibility and delivering the necessary abstraction to e.g. offload or on load operations; The platform's learning services may also publish triggers to which the application components can react in a choreography.

An important research question is how to incentivize application developers to embed their self-adapting capabilities of the IoRT ecosystem. One important consideration is that if applications are "absorbed" in the ecosystem, users might no longer be able to accredit added value to a specific service, which might decrease their willingness to pay (a negative effect for developers).

4.2.10 Machine Learning as Enabler for Adaptive Mechanisms

The IoT community is increasingly experiencing the need to exploit the potential of Machine Learning (ML) methodologies, progressively including them as part of the "things" of the IoT, and contributing to define the contours of a growing need for ML as a distributed service for the IoT. Such a need is mainly motivated by the necessity of making sense of the vast volumes of noisy and heterogeneous streams of sensorial data that can be generated by the nodes in the IoRT, and to approach the challenges posed by its many application domains. Under a general perspective, the convergence between IoT and ML would allow to systematically provide to the IoRT solutions the ability to adapt to changing contexts, at the same time providing high degree of personalization and enabling IoRT applications as well as the very same process management and service organization components of the IoRT architecture to learn from their settings and experience.

The ML service should not only be distributed, whereas it needs allowing embedding intelligence on each node of the IoRT, even at the edge of the network. Such a distributed and embedded intelligence will then be able to perform early data fusion and predictive analyses to generate high-level/aggregated information from low-level data close to where this raw data is produced by the device/sensor or close to where the application consumes the predictions. Such aggregated predictions may, in turn, become an input to another learning model located on a different network node where further predictions and data fusion operations are performed, ultimately constructing an intelligent network of learning models performing incremental aggregations of the sensed data.

Figure 4.3 shows a high-level description of how such a distributed learning architecture maps to a network of intelligent robotic things, highlighting the learning models embedded on the IoRT devices, with different computational, sensing and actuation capabilities (depicted by different colours and sizes in the figure). Figure 4.3 shows how the sizing of the learning models needs to be adjusted to the computational capabilities of the hosting device: some devices might only serve as input data providers for remote learning models. More powerful computing facilities, e.g. cloud services, can be used to deploy larger and more complex learning models, for instance aggregating predictions from several distributed learning models to provide higher-level predictions (e.g. at the level of regional gateways).

Learning service predictions need to be provided through specialized interfaces for applications and IoRT services, implementing different access policies to the learning mechanisms. One of the key functionalities such a service will need to offer, is the possibility of dynamically allocating new predictive learning tasks upon request, and the deployment of the associated learning modules, based on example/historical data supplied by the IoRT

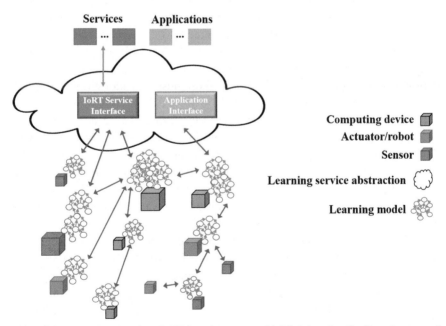

Figure 4.3 Architecture of an IoRT learning system highlighting the distributed nature of the service and the thing-embedded learning models.

applications or the platform services. Altogether such interfaces serve to realize an abstraction (depicted by the cloud in Figure 4.3) for the functionalities of the learning service which hinders the complexity of learning task deployment and execution as well as the distributed nature of the system.

From a scientific perspective, the overarching challenge is how to support applications and platform services in their self-adaptivity throughout distributed machine learning on IoT data. Fundamental challenges regarding interoperability need to be addressed, such as how can applications and services formulate data processing requests for currently missing knowledge and how these are translated into appropriate deployment strategies (What learning model to use? Where to deploy trained learning module?). Resource reasoning is another aspect to be carefully addressed: resource consumption needs to consider when determining the deployment of a trained learning module, or *predictor*, and should be constantly monitored (e.g. to dynamically transfer a predictor if resources are insufficient or critical).

Key scientific challenges also relate to the design of the learning models and machinery at the core of an IoRT learning service. These must be designed to cope with the heterogeneity of the computational resources available in the networks nodes and need to be tailored to the specific nature of the low-level data to be processed and aggregated. The latter typically characterizes as fast-flowing time-series information with widely varying semantics, properties and generation dynamics produced by the heterogeneous sensors deployed in the IoRT environment. Based on these considerations, the family of recurrent neural network models from the Reservoir Computing (RC) [12] paradigm can be thought of as particularly suitable to be considered as a ground for the design of the learning modules in an IoRT learning service. RC networks are characterized by an excellent trade-off between the ability to process noisy sensor streams and a computational and memory fingerprint, which allows their embedding on very low power devices [13]. Besides the great applicative success in approaching a huge variety of problems in the area of temporal sequence processing (see e.g. [14]), here we find particularly relevant to point out that RC models have been the key methodologies for building the Learning Layer system of the EU-FP7 RUBICON project [15], enabling the realization of a distributed intelligent sensor network supporting self-adaptivity and self-organization for robotic ecologies. The approach developed in RUBICON can be seen as a stepping stone upon which to build an IoRT learning service, by extending it to deal with the larger scale, increased complexity and heterogeneity of the IoRT environment with respect to that of a more controlled robotic ecology.

4.2.11 End to End Operation and Information Technologies Safety and Security Framework

At IoRT systems it is a real challenge increasing safety and security and at the same time implement the cooperation between networks of cameras, sensors and robots, which can be used for simple courier services, and also to include information coming from continuously patrol the environment and to check for suspicious/anomalous event patterns, and avoid the multiple possible security breaches.

IoRT End to end services must take into consideration that increasing users' comfort and energy efficiency is required. End to end safety and security services need to enable accounting for groups of users the requirements, remembering them across repeated visits, and seamlessly incorporating them into the building's heating and cooling policies, and by exploiting service robots to provide feedback on energy usage and to ensure that all the sensors in the building are calibrated and in working conditions.

IoRT challenge is to guarantee that the types, amount, and specificity of data gathered by robots and the number of billions of devices creates concerns among individuals about their privacy and among organizations about the confidentiality and integrity of their data. Providers of IoRT enabled products and services should create compelling value propositions for data to be collected and used, provide transparency into what data are used and how they are being used, and ensure that the data are appropriately protected.

IoRT poses a challenge for organizations that gather data from robotic systems and billions of devices that need to be able to protect data from unauthorized access, but they will also need to deal with new categories of risk that the having the Internet of Robotic Things connected to the Internet permanently can introduce. Extending information technology (IT) systems to new devices creates many more opportunities for potential breaches, which must be managed. Furthermore, when IoRT is deployed control of physical assets is required thus the consequences associated with a breach in security extend beyond the unauthorized release of information because potentially cause of the potential physical harm to individuals.

4.2.12 Blockchain

Blockchain technologies, including distributed ledgers and smart contracts, allow IoRT technologies and applications to scale securely, converge, combine and interact across various industrial sectors. The technology enables a decentralised and automated IoT infrastructure that allows trust less decentralized and autonomous applications to interact and exchange data and services. The ability of blockchains and other distributed technologies to enable automated and intelligent machine to machine (robotic things) networks are transforming the design, manufacturing, distribution, logistics, retail, commerce and health applications. This will impact almost every supply chain from health to construction and manufacturing.

Figure 4.4 depicts the distributed ledger technology of blockchain that allows that in each stage of a transaction is generating a set of data, which are called blocks and as the transaction progresses, blocks are added, forming a chain, while encryption software guarantees that the blocks cannot be deleted or changed. Blockchain relies on peer-to-peer agreement (not a central authority) to validate a transaction and the transacting stakeholders rely on an open register, the ledger, to validate the transaction.

The blockchain software is installed on different computing nodes across a network and each transaction is shared to these nodes in the network and the nodes compete to verify the transaction, since the first that verifies, adds the block of data to the chain and gets an incentive, while the other nodes check the transaction, agree on about its correctness, replicate the block, and keep an updated copy of the ledger, as a form of proof that the transaction occurred.

The blockchain integrated into IoRT allows AI-based edge and cloud intelligence solutions for robotic things, using secure low latency communications technology. This allows the training and machine to machine learning

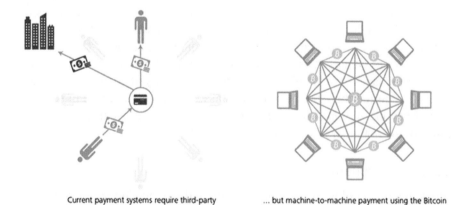

Current payment systems require third-party intermediaries that often charge high processing fees ...

... but machine-to-machine payment using the Bitcoin protocol could allow for direct payment between individuals, as well as support micropayments.

Figure 4.4   Blockchain – Payment process – Current vs Bitcoin [21].

not only one by one but training many robotic things by having edge and cloud intelligence that update in real-time in the field the robotic things with new and improved skills. The extended capabilities can use virtual reality and augmented reality for secure training.

A blockchain-enabled convergence framework is presented in Figure 4.5 to visualise the trends as a cohesive stack. The bottom data collection layer includes any sensor or hardware connected to the Internet receiving and transmitting data. This is essentially the IoT and includes devices, smartphones, drones, autonomous vehicles, 3D printers, augmented and virtual reality headsets, and connected home appliances.

The data is fed into the data management layer, with the role to manage the data being collected and the layer has different components of a decentralised architecture. The specific products can be swapped in and out, using a file system and storage component, a processing and database component and a ledger component.

These components are part of one single platform or best-of-breed for each. The data automation layer uses the data to automate business process and decision making. The automation will come from smart contracts utilizing other data directly from the ledger or smart contracts using oracles to pull data from outside of the system. Artificial narrow intelligence (ANI) can be integrated directly into the smart contract or can be the oracle itself. The higher layer is the organisational structure that directs the activity in the below layers.

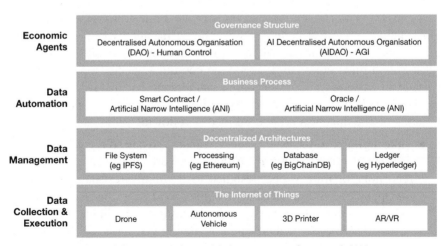

Figure 4.5 Blockchain-Enabled convergence framework [11].

The whole stack can be governed by a decentralised autonomous organisation controlled by human actors, or at some point in the future, the entire stack can be managed by an AI DAO, which may or may not constitute an artificial general intelligence (AGI). Blockchains, artificial intelligence, IoT, autonomous robotics, 3D printing, and virtual and augmented reality are all converging to significantly disrupt existing industries and create whole new markets and economic models [11]. The framework presented need to be integrated as part of the IoT open platforms architecture presented in Section 4.3.

Blockchain-based data marketplace provides a way to share and monetize data and new business models can be created so that data providers can rent their data for a specific experiment, or time period, or even based on outcomes. Autonomous robots are machines that are the mechanical manifestation of artificial intelligence and they use machine learning techniques to make decisions without needing to be pre-programmed.

Blockchain-based data marketplace provides a way to share and monetize data and new business models can be created so that data providers can rent their data for a specific experiment, or time period, or even based on outcomes. Autonomous robots are machines that are the mechanical manifestation of artificial intelligence and they use machine learning techniques to make decisions without needing to be pre-programmed. Deep learning and reinforcement learning are being applied to computer vision and natural language processing problems enabling robots to learn from experience. These sorts of advances are making it possible for robotic things to be used in autonomous vehicles, drones, retail robots applications. The benefits of blockchains or more specifically machine to machine robotics space. As drones and vehicles turn autonomous, they need a way to share and transact data and importantly, in networks, to coordinate decisions. Blockchains provide a way to achieve group consensus more effectively [11].

The blockchain can use to for different purposes as presented in Figure 4.6. The three levels are described as following [63]:

- Store digital records: where blockchain uses advanced cryptography and distributed programming to achieve a secure, transparent, immutable repository of truth – one designed to be highly resistant to outages, manipulation, and unnecessary complexity. In the trust economy, the individual – not a third party – will determine what digital information is recorded in a blockchain, and how that information will be used and the users may record:

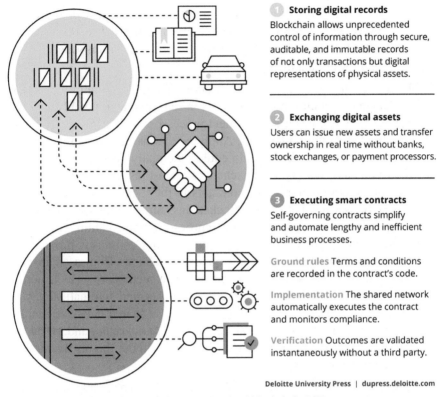

Storing digital records

Blockchain allows unprecedented control of information through secure, auditable, and immutable records of not only transactions but digital representations of physical assets.

Exchanging digital assets

Users can issue new assets and transfer ownership in real time without banks, stock exchanges, or payment processors.

Executing smart contracts

Self-governing contracts simplify and automate lengthy and inefficient business processes.

Ground rules Terms and conditions are recorded in the contract's code.

Implementation The shared network automatically executes the contract and monitors compliance.

Verification Outcomes are validated instantaneously without a third party.

Deloitte University Press | dupress.deloitte.com

Figure 4.6 Three levels of blockchain [63].

- Digitized renderings of traditional identity documents such as driver's licenses, passports, birth certificates, social security/ medicare cards, voter registration, and voting records
- Ownership documents and transactional records for property, vehicles, and other assets of any form
- Financial documents including investments, insurance policies, bank accounts, credit histories, tax filings, and income statements
- Access management codes that provide any identity-restricted location, from website single sign-on to physical buildings, smart vehicles, and ticketed locations such as event venues or airplanes
- A comprehensive view of medical history that includes medical and pharmaceutical records, physician notes, fitness regimens, and medical device usage data

○ As a repository of valuable data, blockchain can provide individual users with control over their digital identities. It can potentially offer businesses an effective way to break down information silos and lower data management costs.

- Exchange digital assets without friction: using blockchain, parties can exchange ownership of digital assets in real time and, notably, without banks, stock exchanges, or payment processors – all applications requiring trusted digital reputations. Applying that basic transactional model to P2P transactions, blockchain could potentially become a vehicle for certifying and clearing asset exchanges almost instantaneously.
- Execute smart contracts: not contracts in the legal sense, but modular, repeatable scripts that extend blockchains' utility from simply keeping a record of financial transaction entries to implementing the terms of multiparty agreements automatically. Using consensus protocols, a computer network develops a sequence of actions from a smart contract's code. This sequence of actions is a method by which parties can agree upon contract terms that will be executed automatically, with reduced risk of error or manipulation. With a shared database running a blockchain protocol, the smart contracts auto-execute, and all parties validate the outcome instantaneously – and without the involvement of a third-party intermediary.

The concept can be used for IoRT applications that exchange information and create collaborative networks among of various fleets of IoRT devices. Swarm robotics is such an application with a strong influence from nature and bio-inspired models and known for their adaptability to different environments and tasks. The fleets of robotic swarms characterised by their robustness to failure and scalability, due to the simple and distributed nature of their coordination [22]. One of the main obstacles to the large-scale deployment of robots for commercial applications is security. The security topic was not properly addressed by state-of-the-art research mainly due to the complex and heterogeneous characteristics of robotic swarm systems – robot autonomy, decentralized control, many members, collective emergent behaviour, etc. Technology such as blockchain can provide not only a reliable peer-to-peer communication channel to swarm's agents, but are also a way to overcome potential threats, vulnerabilities, and attacks. In [22] the blockchain encryption scheme is presented and techniques such as public key and digital signature cryptography are considered accepted means of not only making transactions using unsafe and shared channels, but also of proving the identity of specific agents in a network. A pair of complementary keys, public and private, are created for each agent to provide these capabilities, as presented in Figure 4.7.

Public keys are an agent's main accessible information, are publicly available in the blockchain network, and can be regarded as a special type of account number. Private keys are an agent's secret information, like passwords in traditional systems and are exclusevly used to validate an agent's identity and the operations that it may execute. In the case of IoRT and swarm robotics, public key cryptography allows robots to share their public keys with other robots who want to communicate with them. Any robot in the network can send information to specific robot addresses, knowing that only the robot that possesses the matching private key can read the message. Since the public key cannot be used to decrypt messages, there is no risk if it is intercepted by other robot/person. Public key cryptography prevents third-party robots from decrypting such information even if they share the same communication channel. Digital signature cryptography, as presented in Figure 4.7. allows robots to use their own private key to encrypt messages. The othe IoRT robots can then decrypt them using the sender's public key. All the robots in the fleet have access to the sender's public key, the contents of the message is not a secret, and since it was encrypted using the sender's private key proves that the message could not have been sent by anyone else, thereby proving its authorship. Public key cryptography ensures that the content of a message, encapsulated in a blockchain transaction, can only be read by the robot owning a specific address, while on the other hand, digital signature cryptography provides entity authentication and data origin authentication between robots or third-party agents [22].

4.3 IoRT Platforms Architecture

The IoT developments in the last few years have generated multiple architectures, standards and IoT platforms and created a highly fragmented IoT landscape creating technological silos and solutions that are not interoperable with other IoT platforms and applications. To overcome the fragmentation of vertically-oriented closed systems, architectures and application areas and move towards open systems and platforms that support multiple applications, new concepts are needed for enhancing the architecture of open IoT platforms by adding a distributed topology and integrating new components for integrating evolving sensing, actuating, energy harvesting, networking and interface technologies.

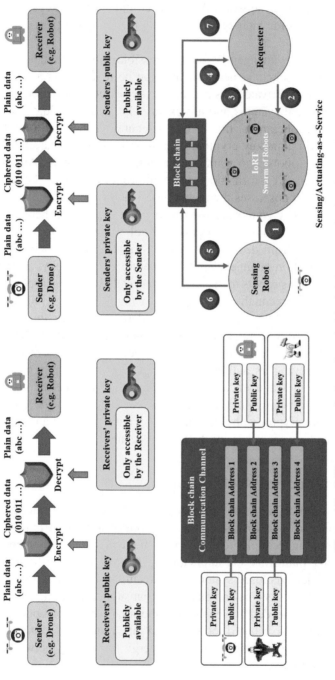

Figure 4.7   Different types of robots share the blockchain communication channel [22].

An IoT Platform can be defined as an intelligent layer that connects the things to the network and abstract applications from the things with the goal to enable the development of services. The IoT platforms achieve several main objectives such as flexibility (being able to deploy things in different contexts), usability (being able to make the user experience easy) and productivity (enabling service creation to improve efficiency, but also enabling new service development). An IoT platform facilitates communication, data flow, device management, and the functionality of applications. The goal is to build IoT applications within an IoT platform framework. The IoT platform allows applications to connect machines, devices, applications, and people to data and control centres. The functionally of IoT platforms covers the digital value chain of an end-to-end IoT system, from sensors/actuators, hardware to connectivity, cloud and applications. IoT platforms' functionalities cover the digital value chain from sensors/actuators, hardware to connectivity, cloud and applications. Hardware connectivity platforms are used for connecting the edge devices and processing the data outside the datacentre (edge computing/fog computing), and program the devices to make decisions on the fly. The key benefits are security, interoperability, scalability and manageability by using advanced data management and analytics from sensor to datacentre. IoT software platforms include the integration of heterogeneous sensors/actuators, various communication protocols abstract all those complexities and present developers with simple APIs to communicate with any sensor over any network. The IoT platforms also assist with data ingestion, storage, and analytics, so developers can focus on building applications and services, which is where the real value lies in IoT. Cloud based IoT platforms are offered by cloud providers to support developers to build IoT solutions on their clouds [5].

The IoT platforms implementations across different industry verticals reveal the use of more than 360 IoT platforms that are using Platform-as-a-Service (PaaS), Infrastructure-as-a-Service (IaaS), Software-as-a-Service (SaaS) deployments. IoT PaaS platforms are built based on event-based architectures and IoT data and provide data analysis capabilities for processing and managing IoT data. IoT-as-a-Service can be built on these different deployments. All the deployments (i.e. SaaS, PaaS and IaaS) have their challenges and security is one important issue that is connected to identity and access management.

Infrastructure as a Service (IaaS) providers and Platform as a Service (PaaS) providers have solutions for IoT developers covering different application areas. PaaS solutions, abstract the underlying network, compute, and storage infrastructure, have focus on mobile and big data functionality, while moving to abstract edge devices (sensors/actuators) and adding features for data ingestion/processing and analytics services [5].

The IoRT applications require holistic multi-layer, multi-dimensional architectural concepts for open IoT platforms integrating evolving sensing, actuating, energy harvesting, networking and interface technologies. This includes end-to-end security in distributed, heterogeneous, dynamic IoT environments by using integrated components for identification, authentication, data protection and prevention against cyber-attacks at the device and system levels, and can help ensure a consistent approach to IoT standardisation processes.

In this context, the IoT platforms need to integrate new components in the different IoT architecture layers to address the challenges for connectivity and intelligence, actuation and control features, linkage to modular and ad-hoc cloud services, data analytics and open APIs and semantic interoperability across use cases and conflict resolution by addressing object identity management, discovery services, virtualisation of objects, devices and infrastructures and trusted IoT approaches.

The IoRT platforms architectures allow robotic things, local embedded and/or distributed intelligence, and smart networks to interact and exhibit smart behaviour and ultimately create open and sustainable marketplaces for large-scale complex and heterogeneous IoT applications and services. Due to the heterogeneity of the applications, devices and stakeholders IoT platforms generic architectures need to be independent of any specific application domains, which refer to the areas of knowledge or activity applied for one specific economic, commercial, social or administrative scope. The architectural concept builds on the common requirements based on use cases of the IoT and the IoT stakeholders, considering key areas from a requirement perspective combined with representative use cases of the IoT that are abstracted from application domains.

The IoT developments in the last few years have generated multiple architectures, standards and IoT platforms and created a highly fragmented IoT landscape creating technological silos and solutions that are not interoperable with other IoT platforms and applications. In order to overcome the fragmentation of vertically-oriented closed systems, architectures and application areas and move towards open systems and platforms that support multiple applications, there is a need for enhancing the architecture of open IoT platforms by adding a distributed topology and integrating new components for integrating evolving sensing, actuating, energy harvesting, networking and interface technologies. The key technological shift is to provide tools and methods for implementing components and mechanisms in different architectural layers that operates across multiple IoT architectures, platforms and applications contexts and add functionalities for actuation and smart behaviour. One solution as presented in the layered architecture concept in Figure 4.8 is that the services and applications are running on top of a specific architectural layer and provide higher-level functionalities such as e.g. data filtering and complex event management and processing that allow the services of existing IoT platforms to be integrated. This concept allows solution providers to use, share, reuse the data streams and perform analytics on shared data increasing the value added of IoT applications.

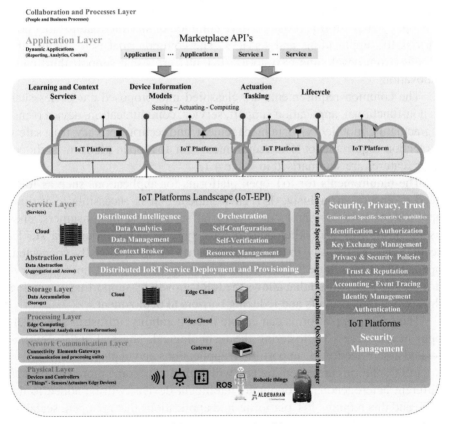

Figure 4.8 IoRT layered architecture.

The IoT applications using this approach integrate data and services among different IoT platforms and between different applications, using shared infrastructure and common standards and reducing the cost for deployment and maintenance. Application developers are able to reuse their applications in different applications, across the IoT ecosystem and greatly reducing development effort and time.

This approach allows to develop a strong IoT ecosystem around the architectural concept providing tools and methods to be used for a number of open IoT platforms that offer solutions across multiple applications and verticals. The ecosystem is built via a combination of tight and loose partnerships between the various industry, and other partners that leads to flexibility in adapting various innovative business models that is demonstrated for heterogeneous systems including autonomous, robotic type of edge devices. The open IoT platforms provided have common or specific features that host various IoT applications and services. The common goal is to capture the benefits from developing easy-to-use IoT platforms that support third party innovation.

The common requirements are classified into proposed categories such as non-functional, application support, service, communication, device (sensing/actuating/mobile/fix), data management, and security, privacy, trust safety protection requirements. The requirements for IoT open platform architectures features are summarised in Table 4.1.

The requirements for IoT open platforms for applications such as IoRT need to ensure an inclusive IoT environment that is accessible to various applications verticals across the industrial sectors and to consumers, end-users, businesses and other autonomous systems. This requires a stable, secure, and trustworthy IoT environment that assure a globally connected, open, and interoperable IoT platforms and environments built upon industry-driven, standards-based that allows the IoRT growth by supporting expanding the applications markets and reducing barriers to deployment.

The IoT open platforms can enable interoperability, infrastructure development and access by fostering the technological, physical and spectrum-related assets needed to support IoRT applications and deployments.

IoRT solutions are emerging and will scale and become more complex as different heterogenous autonomous intelligent devices will be added to the edge and this requires IoT platforms and applications that are open, scalable, extensible, safety and secure.

Table 4.1 IoT open platform architecture requirements

Features	Description
Authentication and authorization	Support multi-layer authentication and authorization
Auto-configuration	Support auto-configuration that allows the IoT system to react to the addition and removal of components such as edge devices and networks.
Autonomous management	Support self-configuring, self-optimizing, self-healing, self-protecting capabilities, for adapting to various application domains, different communication environments, different numbers and types of edge devices.
Compliant components	Support the connection and integration of various heterogeneous set of components performing differing functions based on stakeholders' and applications requirements. Architectural support for discovery and use of components whose characteristics are known and described using standardized semantics and syntaxes.
Cognitive and Artificial Intelligence	Support the cognitive and artificial intelligence components, processes and operations at different IoT architectural layers including end-to-end security.
Privacy and confidentiality	Support for privacy and confidentiality of IoT applications. Possibility to address to scale the solutions and offer context-based implementations.
Content-awareness	Support content-based awareness to enable and facilitate services for path selection and routing of communications, or configuration decisions based on content.
Context-awareness	Support context-based awareness that enable flexible, user-customized and autonomic services based on the related context of IoT components and/or users. The context-based information forms the basis for taking actions in response to the current situation, possibly using sensors and actuators information.
Data analytics	Support for analytics components performed at the different IoT layered architecture, cloud or edge including real-time, batch, predictive, and interactive analytics. The real-time analytics conduct online (on-the-fly) analysis of the streaming data. Batch analytics runs operations on an accumulated set of data. Predictive analytics focusing on making predictions based on various statistical and machine learning techniques. Interactive analytics runs multiple exploratory analysis on both streaming and batch data.

(*Continued*)

Table 4.1   Continued

Features	Description
Data collection protocols	Support for various types of protocols used for data communication between the components of an open IoT platform that need to be scaled to large number of heterogeneous edge devices. Lightweight communication protocols used to enable low energy use as well as low network bandwidth functionality.
Discovery services	Support discovery services across domains and applications for IoT users, services, capabilities, devices and data from devices to be discovered according to different criteria, such as geographic location information, type of device, etc.
Distributed end-to-end security	Support an end-to-end framework for security with secure components, communications, access control to the system and the management services and data security. Physical, digital, virtual and cyber security aspects need to be considered. Support for blockchain components and distributed implementations.
Heterogeneity	Support heterogeneous devices and networks with different types of edge devices regarding communication technology, computing capabilities, storage capability and mobility, different service providers and different users and support interoperability among different networks and operating systems. Support for universal, global-scale connectivity including legacy system interworking.
Location-awareness	Support for IoT components that interact with the physical world and require awareness of physical location, while the accuracy requirement for location is based upon the application. Components describe their locations, and the associated uncertainty of the locations.
Manageability	Support management capabilities to address aspects such as data management, device management, network management, and interface maintenance and alerts. Availability of lists of edge devices connected to the IoT platform, while tracking the operation status, handle configuration, firmware updates, and provide device level error reporting and error handling.
Modularity	Support components that can be combined in different configurations to form various IoT systems. Standardized interfaces for providing flexibility to implementers in the design of components and IoT systems.
Monetization	Support for monetization of functionalities of robots is crucial as an incentive for ecosystem participation. Examples for such monetization range from micro payments for ordering the help of a service robot

Table 4.1 Continued

	at an airport, to ordering a fully customized manufacturing process at an automated plant. Besides the monetization of functionalities and services of robots, the data collected by robots can be monetized as well. For both aspects, functionalities and data, concepts and mechanisms for monetization, such as an ecosystem-wide *marketplace*, are required.
Network connectivity	Support connectivity capabilities, which are independent of specific application domains, and integration of heterogeneous communication technologies needs to be supported to allow interoperability between different IoT devices and services. Networked systems may need to deliver specific Quality of Service (QoS), and support time-aware, location-aware, context-aware and content-aware communications
Openness	Support IoT platforms openness, based on standardised, interoperable solutions allowing any edge device, from any IoT platform, to be able to connect and communicate with one another.
Regulation compliance	Support compliance with relevant application domain specific regulations and regional requirements.
Reliability	Support the appropriate level of reliability for communication, service and data management capabilities to meet system requirements. Provide resilience and support the ability to respond to change due to external perturbations, error detection and self-healing.
Risk management	Support operational resilience under normal, abnormal and extreme conditions.
Scalability	Support a large range of applications varying in size, complexity, and workload. Support systems integrating evolving sensing, actuating, energy harvesting, networking, interface technologies, involving a large number of heterogeneous edge devices, applications, users, significant data traffic volumes, frequencies of event reporting etc. Provisions for components that are used in simple applications to be usable in large-scale complex distributed IoT systems.
Shared vocabularies	To be able to build up ecosystems of robots and IoRT platforms, it is crucial to establish shared vocabularies as a basis for interweaving them and enabling collaboration. Thereby, such shared vocabularies are needed wherever data is serialized and transmitted or exchanged.

(Continued)

Table 4.1   Continued

Features	Description
	The types, terms and concepts in the data (e.g., measured data, metadata, authorization data) need to be defined and these definitions should be part of documented vocabularies so that they can be correctly (re)used.
Standardised interfaces	Support standardised interfaces to the platforms components at different architectural layers based on established, interpretable, and unambiguous standards. Standardized web services for accessing sensors/actuators information, sensors observations and actuators actions.
Support for legacy components	Support legacy component integration and migration, while new components and systems are designed considering that present or legacy aspects do not unnecessarily limit future system evolution. Legacy components integrations need to ensure that security and other essential performance and functional requirements are met.
Time-awareness	Support for event management including time synchronicity among the actions of interconnected components by using communication and service capabilities. Time stamp associated to a time measurement from the physical world and combine or associate data from multiple sensors/actuators and data sources.
Timeliness	Support timeliness, in order to provide services within a specified time for addressing a range of functions at different levels within the IoT system.
Unique identification	Support standardised unique identification for each component of the IoT (e.g. edge devices and services) to provide interoperability, support services (i.e. discovery and authentication across heterogeneous networks) and address object identity management.
Usability	Plug and Play capabilities to enable on-the-fly, on-the-air generation, composition or the acquisition of semantic-based configurations for seamless integration and cooperation of interconnected components with applications, and responsiveness to application requirements.
Virtualisation	Virtualisation of edge objects, networks and layers.

4.3.1 IoRT Open Platforms Architectural Concepts

The heterogeneous IoT devices communicate and transmit data to other devices, gateways and to edge or cloud based IoT platforms where the data is analysed and exchange among applications through systems that take decisions, visualize issues and patterns, steer processes and create new services.

In this, dynamic heterogeneous environment the open platforms architectural concepts play a critical role as there are interactions among intelligent devices across the platforms and application domains.

Figure 4.9 shows the different operations an IoRT open platform should include. The architecture is inspired on the emerging microservices concept, which fosters loose coupling and extendibility.

IoT sensors and actuators post raw or pre-processed data on a distributed event bus, directly or via an IoT gateway. Note that also robots can push sensor observations and actuator statuses. Other services can subscribe to this data, if they are authenticated and authorized. Example services are context creators that semantically enrich the data, IoRT business processes that send actuation tasks to the IoRT platform, and flexible learning services as described in Section 4.2.10.

The IoRT open platform will pick up IoRT tasks from the event bus and perform the necessary reservation and allocation of actuation resources. SLAs and policies govern which tasks can be executed at which time. For example, a robot monitoring task may not be executed in private spaces. All tasks are scheduled by the IoRT platform and translated into concrete actuation plans. These plans may be the result of orchestration mechanisms deciding how multiple things can work together to achieve application objectives and how data should be shared and functions distributed across the system.

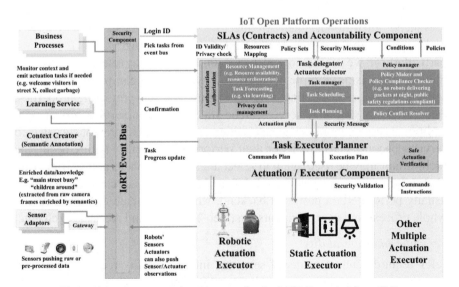

Figure 4.9 A conceptual architecture for the IoRT. Extended from [56].

After the planning stage and validation of policies, the corresponding actuation commands are sent at the appropriate time to the corresponding actuators. Actuators in physical space will perform a final check to see if the requested action is compliant to all safety regulations. This safe actuation verification may be done on the edge, but for robots this is typically implemented as a reactive module on the robot itself.

Task progress is again reported onto the event bus, allowing to adjust plans upon action failures, or to formulate new tasks emerging from the observations made by the robot. The result is a closed and continuous loop.

4.3.2 IoRT Open Platforms Interoperability

Interoperability is one of the topics that has been evolving in the last years with a lot of efforts not only from research communities but industry, the protocols and standards that exist for technical interoperability has been discussed extensively. About semantic interoperability it is common to make use of various IoT standards and platforms that exist today and that can be used for the provisioning of data gathered by smart objects. To enable cross-domain syntactic and semantic interoperability, existing IoT technologies publish open APIs and/or (semantic) data models (e.g., formalized ontologies).

Organisational Interoperability

Advanced Software-Oriented Architecture (SOA) concepts, such as service orchestration and service choreography remains active in IoRT systems, particularly if we are having an increasingly important role in overcoming the ever-increasing complexity of IoRT systems by equipping them with self-configuring, self-healing, self-optimising, and self-protecting properties, etc. (self-*, in short). A service orchestrator acts as a service broker with additional service monitoring capabilities. In cases where previously selected services become unavailable, or their performance drop, or failure occur, the orchestrator may be used again to select alternative services and/or triggering alternative service compositions. IoRT implementations, orchestration will be usually performed on powerful backend, which coordinate and integrate the whole process and its participants via (web-) services and message exchange.

Semantic Interoperability

IoRT requirements in terms of semantic interoperability requires to extend existing ontologies to support the exploitation of robotic elements such as skills, services, shared strategies, and mutual tasks and goals. Further, engineering aspects should be modelled to allow service orchestration distributed over multiple robotic things, also to enable self-* functionalities. This includes describing mutual, context-dependent configurations for resource sharing, negotiation, and conflict resolution among multiple cross-domain services. Advance the concepts around IoT platforms to enable them to provide access to actuation and smart behaviour of robotic things is a possibility, but also the generation of new vocabularies and formalizations around robotic domains. To do this, we can build up on related work, such as existing actuator ontologies (e.g., IoT lite, or the newly published SOSA ontology by W3C). These ontologies already define terms for actuating device and related concepts. However, those ontologies do not go deeper in the modelling of the interrelations of the actuating device. This contrasts with the term of sensing device in those ontologies, which is linked to various other concepts, as it is the traditional focus.

Syntactic Interoperability

IoRT systems requires in term of syntactic interoperability enhance exiting open APIs to enable key functionalities needed by robotic things on IoT platforms, such as discovery, actuation, tasking, and lifecycle management.

One form of syntactic interoperability is computational harvesting, i.e. offloading of computational workload. This has been demonstrated in two ways: First, from a resource-constrained device to an edge cloud. There is challenging energy-performance trade-off between on-board computation and the increased communication cost, while considering network latency. Secondly, self-orchestration on edge clouds is related to the opposite direction, i.e. to shift (computational or storage) workloads from the centralized cloud closer to the endpoints (often the sources of data). This allows to reduce latency of control loops, or to mitigate the ingress bandwidth towards centralized servers.

Platform Interoperability

It remains a challenge to support closed-loop systems where sensor information is analysed and used in-situ, and will be necessary investigating de-centralised architectures to overcome the latency and single-point-of-failure problems associated with centralised ones. The associated interaction style, called choreography, is thought to be a more suitable way to enable a seamless integration of so-called smart items or smart objects within general IoT infrastructures. However, rather than on simple devices, e.g. devices with limited configuration options, choreography relies on agent-like IoT entities, i.e. entities able to execute business logic and decision-making processes, and to interact among each other. A clear disadvantage of such an approach is that, at present, it is very difficult to involve computational constrained devices in the choreography, given their computational, power and network constraints. In addition, choreography opens the question of what protocols should be implemented by the smart entities, as no standards yet exist.

In the context of interoperability is important to mention the work of the newly formed IEEE-RAS Working Group, named Ontologies for Robotics and Automation. The group addresses a core ontology that encompasses a set of terms commonly used in Robotics and Automation along with the methodology adopted.

The work uses ISO/FDIS 8373 standard developed by the ISO/TC184/SC2 Working Group as a reference. The standard defines, in natural language, some generic terms which are common in Robotics and Automation such as robot, robotic device, etc. [30]. Several ontologies have been proposed for several robotics subdomains or applications, e.g., search and rescue, autonomous driving, industrial, medical and personal/service robotics. In the domain of autonomous robots, ontologies have been applied [30]:

- To describe the robot environment. A critical competence for autonomous robots is to be able to create a precise and detailed characterization of the environment as individual robot knowledge or as a central shared repository of the objects in it or the location they are moving;
- For the description and/or reasoning about actions and tasks. Autonomous robots are faced with complex, real-time tasks which might require a large amount of knowledge to be stored and accessed. Ontologies have been applied to the structuring of this knowledge and its different levels of abstraction, to describe task-oriented concepts, as metaknowledge for learning methods and heuristics or to define concepts related to actions, actors and policies to constraint behaviour;
- For the reuse of domain knowledge. Ontologies have been used to define robots as objects by describing its structural, functional and behavioural features or to characterize the domain and subdomains of robotics.

A robot is an agent and agents can form social groups, so robots can also form what we call robot groups. The work in [31] present an upper level ontology called Suggested Upper Merged Ontology (SUMO) that has been proposed as a starter document by The Standard Upper Ontology Working Group, an IEEE-sanctioned working group of collaborators from the fields of engineering, philosophy and information science. SUMO provides definitions for general-purpose terms and acts as a foundation for more specific domain ontologies. According to SUMO, a group is "a collection of agents", like a pack of animals, a society or an organization. In this context, a group is an agent, in the sense that it can act on its own. Similarly, to semi-autonomous and non-autonomous robots, the agents that compose the group form their agency. Examples of robot groups are robot teams, such as robot football teams and a team of soldering robots in a factory. This category also encloses what are called complex robots. These are embodied mechanisms formed by many agents attached to each other; e.g., a robotic tank in which the hull and the turret are independent autonomous robots that can coordinate their actions to achieve a common goal. Robots and other devices can form systems. In accordance with ISO, a robotic system is an entity formed by robots (e.g. single robots or groups of robots) and a series of devices intended to help the robots to carry on their tasks (Figure 4.10. Robotic system and its relations with robot and robotic environment [30]). A robotic environment is an environment equipped with a robotic system. Other example of robotic system is an automated home assistant system composed of a helper robot as well as by sensors and actuators to open doors [30].

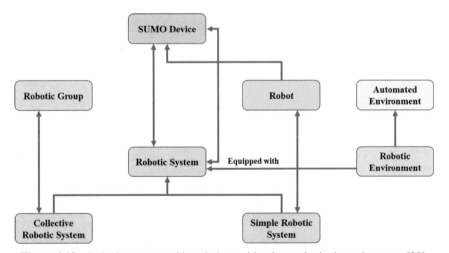

Figure 4.10   Robotic system and its relations with robot and robotic environment [30].

4.3.3 Marketplace for an IoRT Ecosystem

To give incentives for participation and growing of an IoRT ecosystem, mechanisms for monetization of service functions and data are required (Table 4.1). A marketplace needs to be established as a centrepiece of an IoRT ecosystem. Thereby, a marketplace allows the registration and discovery of offerings, i.e., data or functions offered by services. Such services can be standalone components, can be provided by IoRT platforms, or running on a thing or robot itself. The marketplace acts as an exchange point for providers and consumers of offerings. As shown in Figure 4.11, a consumer of offerings is e.g. an application or a service. A provider of an offering is a platform or a service that adds value to an offering of a platform.

According to [62] a marketplace for such ecosystems should provide mechanisms for:

- Registration of offerings, i.e., a provider of an offering can upload a metadata description that is ingested by the marketplace and indexed to support discovery.
- Discovery of offerings, i.e., a consumer utilizes an interface of the marketplace to search for offerings. For registration and discovery, it is crucial to have shared vocabularies for the metadata description of
- Authentication and authorization, i.e., consumer and provider can securely access the marketplace and use role and privilege management (e.g., association with a user group) can be conducted.

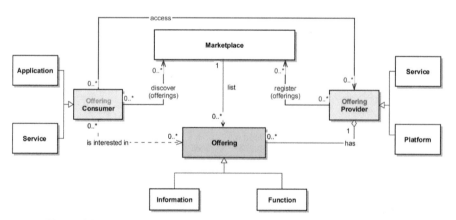

Figure 4.11   Conceptual model of a marketplace for an IoRT ecosystem [61].

- Reputation management, i.e., consumers can rate providers and their offerings; these ratings can be incorporated into search rankings and during discovery.
- Accounting and charging, i.e., the usage of an offering by the consumer is accounted (e.g., API calls are counted) and providers can charge for this usage accordingly. This is a crucial functionality to enable monetization of IoRT offerings and to give incentive for the ecosystem to grow. It is closely related to handling different licenses of data offerings.
- Orchestration, i.e., supporting the design, instantiation, control and sharing of offering compositions. This is not a mandatory functionality of a marketplace; however, it fosters reuse of registered offerings as they will be utilized in multiple workflows. Orchestration can even allow engineering of IoRT applications, e.g., a custom manufacturing process can be modelled as a collaboration of various robotic thing functions.

4.4 IoRT Applications

4.4.1 Introduction

The lessons learnt in researching network robot systems [20], ubiquitous robotics [23] and robotic ecologies [24], is that, although robots are becoming increasingly more autonomous, they are simply more efficient and intrinsically more effective if they are part of ambient intelligence solutions as a natural conditional to have integrated IoT deployed systems with Robotic systems. Patents for robotics and autonomous systems have swelled in the last decade. It is estimated that more than $67 billion will be spent worldwide in the robotics sector by 2025, compared to only $11 billion in 2005, reaching the compound annual growth rate (CAGR) of 9% [25]. Besides robots employed in industry and factory automation, service robotics for use in domestic, personal, and healthcare settings is the fastest growing sector. The World Robot Report projecting sales of 333,200 new robots in the period 2016–19 representing a global market more than 23 Billion US dollars. Integrated IoT & Robotics solutions will increasingly represent a significant proportion of this market. The following sections give a brief overview of opportunities in selected application domains.

Research interest in service robotics for assistance and wellbeing has grown during the last few decades, particularly as consequence of demographic changes. Maintaining a healthy lifestyle and trying to achieve a state of well-being helps to improve the life conditions and increase its durability. Service robotics could focus on early diagnosis and detection of risks, to develop prevention programs. Thus, it is possible to use robots to perform physical activity at home, or planning a proper nutrition program, based on the user's needs. Personal wellbeing management robots can provide services also for people who are alone, or live isolated from families. These robots can both detect physiological parameters and transmit them to the doctor in real time and to interpret the emotional state of the user and accordingly interact. Figure 4.12 illustrate the evolution of robots in different application areas presented as report from Yole Development in 2016 [84].

4.4.2 Predictive and Preventive Maintenance

Machine maintenance for robots and IoT equipment is quite expensive because the dedicated equipment and the necessary to execute that. For instance, maintaining certain equipment may include a "preventive maintenance checklist" which includes small checks that can significantly extend service life. All this information need to be processed by the maintenance robot in real time or at least in the few minutes before the maintenance is

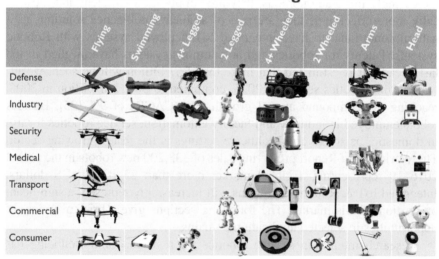

Figure 4.12   Robots classification per application areas and mobility evolution [84].

scheduled to assess the best conditions to perform the maintenance. Multiple external factors, such as weather and equipment are considered; for example, heating systems maintenance is often recommended to be performed before the winter time to prevent failures likewise HVAC is better recommended to be performed before the hottest time of the year.

IoRT treats machine failures as part of the device extension and robots' operation, considering that failures as an inherent characteristic that is generated by the natural degradation of mechanical materials or the silicon degradation suffered as consequence of bringing the modification and operation of the devices and systems. The primary goal of maintenance is to reduce or mitigate the consequences of failure of the devices and the systems associated in their operation and or the equipment around them. IoRT not only look at preventing the failure before it occurs but ideally defines planned maintenance schemes and conditions based on maintenance that will help to achieve certain levels of good operation. Robots usually are designed to preserve and restore equipment offering reliability by indicating clearly what are the parts that are required to be replace and likewise identifying those worn components before they fail. Maintenance includes preventive (partial or complete) overhauls at specified periods, as per example, cleaning, lubrication, oil changes, parts replacement, tune ups and adjustments, and so on. In addition, calibration can be also considering part of the maintenance, workers usually record equipment deterioration so they know to replace or repair worn parts before they cause system failure. IoRT should take care of these conditions and even beyond that the ideal IoRT machine maintenance program would prevent any unnecessary and costly repairs.

4.4.3 Autonomous Manufacturing

According to the International Federation of Robotics (IFR), by 2019, more than 1.4 million new industrial robots will be installed in factories around the world [47]. It is projected that the number of industrial robots deployed worldwide will increase to around 2.6 million units by 2019. Broken down per sectors, around 70 percent of industrial robots are currently at work in the automotive, electrical/electronic and metal and machinery industry segments. While the acquisition costs for such robots are continuously decreasing, the costs for programming them for their specific tasks and environments are still very high. For the future, researchers are working on ways to reduce these costs for programming industrial robots, particularly, by making them more and more autonomous through increased intelligence. i.e., the aim is that we will not specify *how* a robot does something, but we will tell the robot a goal of *what* it should do. Through technologies such as artificial intelligence, the robots will autonomously find a way of how to realize a defined goal.

In this context of increasing autonomy, technologies such as IoT & cloud infrastructure can be used to collect, analyse and visualise real-time production performance indicators, usually to inform existing optimization processes [27], while results from multi-agent systems, and adaptive middleware, can provide advanced suitable coordination and communication protocols to coordinate the operations of multiple robots. Crucial will be in the future the ability of robots to interact and collaborate with human co-workers and ultimately learn from these co-workers on how to conduct a task. Hence, an important topic is to make the co-working of robots and humans in the manufacturing process safer to enable its intensification. Therefore, robots have to be enabled to anticipate human behaviour, while working with them. For instance, Michalos et al., [28] have developed a flexible integration and distributed communication system for data sharing and coordination of autonomous and human-robot collaborative operations, using ontology services to network all possible resources and link them all for higher level coordination by a centralized task planner. Järvenpää et al. [29] have framed production lines as multi-agent systems of heterogeneous devices equipped with self-descriptive capabilities and standardized communication protocols, which they use to negotiate with one another to reduce set-up and changeover times, costs and energy consumption.

4.4.4 Autonomous Logistics, Delivery, e-commerce and Warehouse Automation

The applications in warehouse robotics for IoRT come in response to the rise of e-commerce, where collaborative robots, work alongside human warehouse worker. Logistics firms can use collaborative robots should to ease some of the workforce shortages, and make the work less physically demanding. Delivery using self-driving robots is one typical application for IoRT with fleets of robots, which are designed to operate on pedestrian side and make deliveries within 3–5 kms radius, carrying loads weighing as much as 10 kgs, at speeds of up to 8–10 km/h. The robotic fleets can be monitored remotely and standing by to drive the vehicles remotely if the robots encounter situations are not able to perform in autonomous driving mode.

Amazon has formed a team to investigate how the company might use self-driving technology within its growing logistics network. The team does not intend to design a self-driving vehicle instead it will function as a think tank tasked with helping the e-commerce titan integrate automation into its logistics strategy. The company could use self-driving forklifts, trucks, and other vehicles to expand on its early automation efforts. By further automating logistics, Amazon may be able to cut delivery costs, giving it a key competitive advantage. For example, autonomous forklifts could bring down labor costs in the company's warehouses – the Kiva robots have already cut warehouse operating costs by 20% [46].

4.4.5 Autonomous Home Appliances, and Personal Robots

Personal robots mainly refer to the consumer robotics industry and include solutions to provide services to individuals in personal and household applications.

They are likely to be mass-produced and bought or leased by untrained, or minimally trained people in everyday unstructured environments. The global personal robots market is expected to reach $34.1 billion by 2022 [74]. Typical applications of personal robots concern domestic appliances, telepresence, entertainment, education, and assistance [75].

Domestic environments represent a major place to integrate new technology; several domestic service robots have been introduced as consumer products for the household chores, with a various portfolio of floor-cleaning robots, lawn-mowing robots, security robots, cat litter box robots, decluttering robots, etc. [76].

Telepresence robotics combines communication technology with robots' perception abilities, thus allowing advanced interaction capabilities of humans with remote environments. It allows people to monitor patients or elderly people at home or in hospitals, to virtually move and inspect through distant environments, to participate in work meetings, etc.

Numerous research studies suggest that robotics integration for educational purposes is an effective teaching method, that allow the development of student higher-order thinking skills such as application, synthesis, and evaluation, as well as teamwork, problem solving, decision making, and scientific investigation. Moreover, robotics employed as educational tool help students develop the knowledge and skills required in order to survive in the ever-changing, interconnected Information society era of the 21st century [77].

Cultural heritages, cinemas and retail environment represent a novel and interesting place to integrate new technology. Public and outdoor environments, as a place for technology, are going to have more and more attentions in the future, mainly because a normal life involve the ability to move and live in social and outdoor environments. The panorama of Service Robotics in social activities is wide: visiting cultural heritage, retail environments, outdoor cleaning robots, shopper assistant robots.

Research interest in service robotics for assistance and wellbeing has grown during the last few decades, particularly as consequence of demographic changes. Maintaining a healthy lifestyle and trying to achieve a state of well-being helps to improve the life conditions and increase its durability. Service robotics could focus on early diagnosis and detection of risks, to develop prevention programs. Thus, it is possible to use robots to perform physical activity at home, or planning a proper nutrition program, based on the user's needs. Personal wellbeing management robots can provide services also for people who are alone, or live isolated from families. These robots are able to both detect physiological parameters and transmit them to the doctor in real time and to interpret the emotional state of the user and accordingly interact.

Personal robots represent a new generation of robots that will safely act and interact in the real world of complex environments, and with relatively limited energy consumption and computational resources.

4.4.6 Healthcare Assistants, Elderly Assistance

The value of the healthcare market is significant and there is a key shortage of support provision on a one to one basis for the ageing population. The 'care deficit' poses a major challenge to ageing societies, especially in the EU and Japan. Since care responsibilities towards dependent adults are unpredictable in both duration and intensity of need, greater flexibility is desirable to allow carers to spread their leave or change their working hours to accommodate their changing needs and those of their dependants.

Autonomous and interactive robots integrated with smart environments for Ambient Assisted Living (AAL) applications have been demonstrated in several research projects [35]. On one hand, the smart environment can act as a service provider for the robot, e.g. feeding it with information about the user's whereabouts and state, by using sensors pervasively embedded in the environment and/or worn by the user. The robot can then provide useful services thanks to its physical presence and mobility capabilities.

On the other hand, the robot provides the user with a user interface that acts as a personalised representative of the services that the intelligent environment offers. This has been shown to increase the user's acceptance of the technology [33] and offer added value with services such as cognitive stimulation, therapy management, social inclusion/connectedness, coaching, fall handling, and memory aid.

Combining IoT with AI and robotic components to deliver practical, modular, autonomous and self-adaptive IoRT systems has thus the potential to complement and improve the effectiveness of existing care practices by providing automated, continuous assessment of users' conditions and support both self-care and assistive services that can be constantly in tune with users' requirements [15]. One example is the use of humanoid robots in the dementia ward of an elderly care home. Using wearables and environmental sensors, behavioural disturbances like shouting and wandering are detected and used as trigger to send a robot to start a personal intervention to temporarily distract the resident. Meanwhile, a nurse or another caregiver is alerted. The type of intervention (e.g. dialogue, music playing) is also based on context information provided by the IoT [57]. This is a clear example of an IoRT system supporting caregivers. Consumers have also a growing interest to maintain the health and wellbeing through personalized coaching. Personal, companion robots with language natural interactions and other social skills can be used to this effect. The health coaching market is estimated to be a 700-million-dollar business in the USA, $2 billion business worldwide, with an annual growth rate of 18%. IoRT solution have the potential for a large ROI in terms of not only economic factors but also in terms of improving health and well-being of an ageing demographic at a population level. The so-called "silver market" (people aged 55 and older) represent a market of approximately 1500 billion euro per year in EU27, and they spend more on health-related products and household support devices than people on average [36]. This trend is set to become a major lead market for many commercial sectors. Merrill Lynch estimates the value of the Silver Economy at $7 trillion per year worldwide, which makes it the 3rd largest economy in the world. In the past 20 years, consumer spending among those aged 60 and over rose 50% faster compared to those under 30 (Source: Eurostat). Smart homes and robotic solutions supporting independent living and wellness are among the applications domain that can be empowered by adopting IoRT-driven solution. They are also those that expect to benefit the most from the Silver Economy. If telehealth and telecare were scaled up across Europe to reach 10–20% coverage of the population affected by chronic diseases or old age, this could generate potential markets for new products and services in the range of €10–20 billion a year [39].

4.4.7 Cleaning Robotic Things, Cleaning and Inspection Appliances

The IoRT application area with potential for further grow is the service robots for inspection, cleaning and maintenance. In these applications drones, can be used in conjunction with sensors mounted on hard to reach places, such as wind turbines or high-voltage transmission lines. Service fleets of robots are used in specific dangerous, monotonous or unreasonable jobs for humans. Examples are pipe inspections and cleaning, sewer system inspection that detects and map damage highly precisely and facade cleaning robots. Other examples include autonomous robotic systems that enable safe and cost-effective underwater cleaning and inspection of bridge substructures.

The robotic things can be used for various cleaning and surface preparation devices i.e. water jetting, power tools for rust and paint removal or vacuum suction systems.

4.4.8 Buildings, Garden, City Maintenance

A robot on city streets for executing hard work under stress conditions for humans is a perfect use case that would improve conditions in the city, likewise for working on times were in a town street there is no possibility to make people work. This results in a condition for the robot where the sensors in the city are the guide pointers for its operation (additional to its own navigation and sensors systems). Initially, the robot would become to be part one more element of the equipment (infrastructure) of the city and when it is right be more a dynamic element for the citizens, for example a garbage collection robot in times of extreme hot or low temperatures can execute cleaning operations on urban areas while during the traffic times can serve as traffic indicator in front of the vehicle indicating better routes for circulation. However, over time the city sensors and robots should have the capacity to learn to correlate "robot blocked street" and "dirty street" thus decision must be taken on what are the priorities for the robot and/or which is his primary role in the city and select with an event "vehicles jammed in a traffic zone", and adjusts the garbage collection actuation strategy accordingly. Depending on what sensors and actuators are available, the "garbage collection failures" could be correlated with even more indirect events, such as "automatic adjusted roasters only between 10:00AM–11:00AM for example. Note that this is just an example of two situations in city but at home similar activities can be defined, like gardening the back of the house or clean the front before a delivery of a parcel is expected and not after.

4.4.9 Entertainment and Well-Being

Telepresence robotics combines communication technology with robots' perception abilities, thus allowing advanced interaction capabilities of humans with remote environments. It allows people to monitor patients or elderly people at home or in hospitals, to virtually move and inspect through distant environments, to participate in work meetings, etc.

Cultural heritages, cinemas and retail environment represent a novel and interesting place to integrate new technology. Public and outdoor environments, as a place for technology, are going to have more and more attentions in the future, mainly because a normal life involve the ability to move and live in social and outdoor environments. The panorama of Service Robotics in social activities is wide: visiting cultural heritage, retail environments, outdoor cleaning robots, shopper assistant robots.

4.5 Robotics and IoT Multi Annual Roadmap

The interested reader is referred to the 2020 Robotics Multi Annual Roadmap (MAR) [83] for more details on prime opportunities for robotic technology and to SRIA for IoT technologies [1]. The Robotics Multi Annual Roadmap is a technical guide that identifies expected progress within the Robotic community and provides and analysis of medium to long term research and innovation goals and their expected impact.

The MAR recognises that new automation concepts such as IoT and Cyber-Physical Systems (CPS) have the potential to impact and revolutionise the innovation landscape in many application domains, including:

- Precision farming domain, where improvements in the interoperability and communication both between machines working on the farm and to organisations outside of the farm would allow improvements in the processing of harvested crops, efficient transport and faster time to market.
- Civil domain, where many applications for robotics technology exist within the services provided by national and local government. These range from support for the civil infrastructure, roads, sewers, public

buildings, rivers, rubbish collection etc. to support for law enforcement and the emergency services.

- Environmental monitoring domain, where the ability of robotics technology to provide multi-modal data accurately mapped to terrain data has the potential to accelerate the development and deployment of such systems and enhance those services that rely on this data.
- Inspection, maintenance and cleaning domain, where robots' advantage is in their ability to operate continuously in hazardous, harsh and dirty environments. This include drones, which can be used in conjunction with sensors mounted on hard to reach places, such as wind turbines or high-voltage transmission lines.
- Logistic and transport domains, where robots can provide key services including receiving goods, handling material, e.g. within manufacturing sites (intra-logistics), sorting and storage (warehousing), order picking and packing (distribution centres), aggregation and consolidation of loads, shipping and transportation, e.g. in last mile delivery applications.

4.6 Discussion

As the first ICT revolution (from the personal computer, to the internet, to the smartphone and wearable computers, to Cloud and internet of things) has qualitatively augmented the capability to manage data, the personal robot technology will enable a similar dramatic leap in their capability of acting in the physical world. A crucial role will be played by the integration of robots with Internet of Things and Artificial Intelligence. IoT has features of reconnect with different entities like apps, devices and people interaction, which gives the best solution for many application domains. Combination of Robotics, IoT and Artificial Intelligence results in robots with higher capability to perform more complex tasks, autonomously or cooperating with humans. With IoT platform, multiple robots can get easily interconnected between them and with objects and humans, facilitating the ability to transfer data to them without human to computer or humans to humans interaction. Reasoning capabilities coming from the use of machine learning, also exploiting cloud resources [78], for example, brings beneficial effects in terms of system efficiency and dependability, as well as safety for the user, and adaptive physical and behavioural human-robot interaction/ collaboration.

References

[1] O. Vermesan and P. Friess (Eds.). Digitising the Industry Internet of Things Connecting the Physical, Digital and Virtual Worlds, ISBN: 978-87-93379-81-7, River Publishers, Gistrup, 2016.

[2] O. Vermesan and P. Friess (Eds.). Building the Hyperconnected Society – IoT Research and Innovation Value Chains, Ecosystems and Markets, ISBN: 978-87-93237-99-5, River Publishers, Gistrup, 2015.

[3] O. Vermesan, P. Friess, P. Guillemin, S. Gusmeroli, et al., "Internet of Things Strategic Research Agenda", Chapter 2 in Internet of Things – Global Technological and Societal Trends, River Publishers, 2011, ISBN 978-87-92329-67-7.

[4] O. Vermesan, R. Bahr, A. Gluhak, F. Boesenberg, A. Hoeer and M. Osella, "IoT Business Models Framework", 2016, online at http://www. internet-of-things-research.eu/pdf/D02_01_WP02_H2020_UNIFY-IoT_ Final.pdf

[5] A. Gluhak, O. Vermesan, R. Bahr, F. Clari, T. Macchia, M. T. Delgado, A. Hoeer, F. Boesenberg, M. Senigalliesi and V. Barchetti, "Report on IoT platform activities", 2016, online at http://www.internet-of-things-research.eu/pdf/D03_01_WP03_H2020_UNIFY-IoT_Final.pdf.

[6] The Internet of Robotic Things, ABIresearch, AN-1818, online at https://www.abiresearch.com/market-research/product/1019712-the-internet-of-robotic-things/

[7] J. Wan, S. Tang, H. Yan, D. Li, S. Wang, and A. V. Vasilakos, "Cloud robotics: Current status and open issues," IEEE Access, vol. 4, pp. 2797–2807, 2016.

[8] L. Riazuelo et al., "RoboEarth semantic mapping: A cloud enabled knowledge-based approach," IEEE Trans. Autom. Sci. Eng., vol. 12, no. 2, pp. 432–443, Apr. 2015.

[9] G. Hu, W. P. Tay, and Y. Wen, "Cloud robotics: Architecture, challenges and applications," IEEE Netw., vol. 26, no. 3, pp. 21–28, May/Jun. 2012.

[10] IEEE Society of Robotics and Automation's Technical Committee on Networked Robots. online at http://www.ieee-ras.org/technical-commit tees/117-technical-committees/networked-robots/146-networked-robots

[11] Outlier Ventures Research, Blockchain-Enabled Convergence – Understanding The Web 3.0 Economy, online at https://gallery.mailchimp.com/ 65ae955d98e06dbd6fc737bf7/files/Blockchain_Enabled_Convergence. 01.pdf

[12] M. Lukoševićius and H. Jaeger. "Reservoir computing approaches to recurrent neural network training." Computer Science Review, vol. 3(3), pp. 127–149, 2009.

[13] D. Bacciu, P. Barsocchi, S. Chessa, C. Gallicchio, A. Micheli, "An experimental characterization of reservoir computing in ambient assisted living applications", Neural Computing and Applications, vol. 24(6), pp. 1451–1464, 2014.

[14] M. Lukoševićius, H. Jaeger, B. Schrauwen, "Reservoir Computing Trends", KI – Künstliche Intelligenz, vol. 26(4), pp. 365–371, 2012.

[15] M. Dragone, G. Amato, D. Bacciu, S. Chessa, S. Coleman, M. Di Rocco, C. Gallicchio, C. Gennaro, H. Lozano, L. Maguire, M. McGinnity, A. Micheli, G. M.P. O'Hare, A. Renteria, A. Saffiotti, C. Vairo, P. Vance, "A Cognitive Robotic Ecology Approach to Self-Configuring and Evolving AAL Systems", Engineering Applications of Artificial Intelligence, Elsevier, vol. 45, pp. 269–280, 2015.

[16] G.S. Sukhatme, and Matarić, M.J., Embedding Robots Into the Internet, *Communications of the ACM*, 43(5) special issue on *Embedding the Internet*, D. Estrin, R. Govindan, and J. Heidemann, eds. May, pp. 67–73, 2000

[17] A. B., Craig, Understanding Augmented Reality – Concepts and Applications, 1st Edition, ISBN: 9780240824086, Morgan Kaufmann, 2013.

[18] M. Gianni, F. Ferri and F. Pirri ARE: Augmented Reality Environment for Mobile Robots. In: A. Natraj, S. Cameron, C. Melhuish, and M. Witkowski (eds) Towards Autonomous Robotic Systems. TAROS 2013. Lecture Notes in Computer Science, vol. 8069, Springer, Berlin, Heidelberg, 2014.

[19] B.-K. Shim, K.-W. Kang, W-S. Lee2, J-B. Won, and S-H. Han, An Intelligent Control of Mobile Robot Based on Voice Command, 2012 12th International Conference on Control, Automation and Systems, Oct. 17–21, 2012 in ICC, Jeju Island, Korea, 2012.

[20] IEEE RAS Technical Committee on Networked Robots, 6/10/2017, online at http://networked-robots.cs.umn.edu/

[21] Bitcoin: Beyond money, Deloitte University Press, DUPress.com, online at https://dupress.deloitte.com/content/dam/dup-us-en/articles/bitcoin-fact-fiction-future/DUP_847_BitcoinFactFictionFuture.pdf

[22] E. C. Ferrer, The blockchain: a new framework for robotic swarm systems, Cornell University Library, August 2016, online at https://arxiv.org/pdf/1608.00695.pdf

[23] J. Kim, Y. Kim, and K. Lee. The third generation of robotics: Ubiquitous robot. In Proc of the 2nd Int Conf on Autonomous Robots and Agents (ICARA), Emerston North, New Zealand, 2004.

[24] A. Saffiotti, M. Broxvall, M. Gritti, K. LeBlanc, R. Lundh, J. Rashid, B. Seo, and Young-Jo Cho. "The PEIS-ecology project: vision and results." In Intelligent Robots and Systems, 2008. IROS 2008. IEEE/RSJ International Conference on, pp. 2329–2335. IEEE, 2008.

[25] Gitta Rohling, Robots: Building New Business Models, 2017, online at https://www.siemens.com/innovation/en/home/pictures-of-the-future/digitalization-and-software/autonomous-systems-facts-and-fore casts.html

[26] World Robotics Report 2016, 2016, online at https://ifr.org/news/ifr-press-release/world-robotics-report-2016-832/

[27] ESOCC 2016, Vienna, "The HORSE Project: IoT and Cloud Solutions for Dynamic Manufacturing Processes" presentation & short paper, 5 September 2016.

[28] G. Michalos, S. Makris, J. Spiliotopoulos, I. Misios, P. Tsarouchi, G. Chryssolouris, "ROBO-PARTNER: Seamless Human-Robot Cooperation for Intelligent, Flexible and Safe Operations in the Assembly Factories of the Future", (CATS 2014) 5th CIPR Conference on Assembly Technologies and Systems, 13–14 November, Dresden, Germany, pp. 71–76 (2014).

[29] E., Järvenpää, N. Siltala, and M. Lanz, Formal Resource and Capability Descriptions Supporting Rapid Reconfiguration of Assembly Systems. Proceedings of the 12th Conference on Automation Science and Engineering, and International Symposium on Assembly and Manufacturing August 21–24, 2016. The Worthington Renaissance Hotel, Fort Worth, TX, USA.

[30] E. Prestes, et al., Towards a core ontology for robotics and automation, Robotics and Autonomous Systems 61 pp. 1193–1204, Elsevier Press, 2013.

[31] I. Niles, A. Pease, Towards a standard upper ontology, in: Proceedings of the international conference on Formal Ontology in Information Systems – Volume 2001, FOIS'01, ACM, New York, NY, USA, 2001, pp. 2–9.

[32] National Audit Office (2010) Economic model to assess the financial impacts of the Enriched Opportunities Programme for people with dementia in an extra-care housing setting. London: National Audit Office.

[33] G. Cortellessa, et al. A cross-cultural evaluation of domestic assistive robots. In Proceedings of the AAAI Fall Symposium on AI and Eldercare,Washington, DC (USA), 2008.

[34] M. Waibel, M.l Beetz, J. Civera, R. d'Andrea, J. Elfring, D. Galvez-Lopez, K. Häussermann, R. Janssen, J.M.M. Montiel, A. Perzylo, B. Schiessle, M.Tenorth, O. Zweigle and M.J.G. René Van de Molengraft. *RoboEarth – A World Wide Web for Robots.* In Robotics & Automation Magazine, IEEE, vol. 18, no. 2, pp. 69–82, June 2011. doi: 10.1109/MRA.2011.941632.

[35] M. Dragone, J. Saunders, K. Dautenhahn: On the Integration of Adaptive and Interactive Robotic Smart Spaces. Paladyn 6(1), 2015.

[36] Financial Times "Silver Economy Series" 3 November 2014 (http://www.ft.com/intl/topics/themes/Ageing_populations).

[37] Bank of America Merrill Lynch report 2014: The silver dollar – longevity revolution.

[38] Strategic Intelligence Monitor on Personal Health Systems, Phase 2 – Impact Assessment Final Report, 2012, http://ftp.jrc.es/EURdoc/JRC7 1183.pdf

[39] Organisation for Economic Cooperation and Development (OECD), The wellbeing of nations: the role of human and social capital, 2001, online: The wellbeing of nations: the role of human and social capital, 2017.

[40] L., Parker, and Tang, F. 2006. Building multirobot coalitions through automated task solution synthesis. Proc. of the IEEE 94(7): 1289–1305.

[41] R. Lundh, Karlsson, L.; and Saffiotti, A. 2007. Dynamic self-configuration of an ecology of robots. In Intelligent Robots and Systems, 2007. IROS 2007. IEEE/RSJ International Conference on, 3403–3409.

[42] M. Di Rocco, et al. A planner for ambient assisted living: From high-level reasoning to low-level robot execution and back. AAAI Spring Symposium Series, 2014.

[43] FP7 EU project RUBICON (Robotic Ubiquitous Cognitive Network), online http://fp7rubicon.eu

[44] FP7 EU project RobotERA, online, http://www.robot-era.eu/

[45] rtSOA – A Data Driven, Real Time SOA Architecture for Industrial Manufacturing, online at http://www-db.in.tum.de/research/projects/ rtSOA/

[46] J. Camhi and S. Pandolph, Amazon looks to further logistics automation, Web article, online at http://www.businessinsider.com/amazon-looks-to-further-logistics-automation-2017-4?r=US&IR=T&IR=T

[47] Executive Summary World Robotics 2016 Industrial Robots, online at https://ifr.org/img/uploads/Executive_Summary_WR_Industrial_Robots_20161.pdf

[48] K., Kumar et al. 2010. Cloud Computing for Mobile Users: Can Offloading Computation Save Energy? IEEE Computer vol. 43(4), April 2010.

[49] K, Ha, et al. 2014. Towards Wearable Cognitive Assistance, Proc. Of MobiSys, 2014.

[50] Verbelen T., et al. 2012. AIOLOS: Middleware for improving mobile application performance through cyber foraging. Journal of Systems and Sofware, vol. 85(11), 2012.

[51] Bohez S. et al. 2014 Enabling component-based mobile cloud computing with the AIOLOS middleware. Proc of the 13th Wor. on Adaptive and Reflective Middleware, ACM 2014.

[52] De Coninck., et al. 2016. Middleware platform for distributed applications incorporating robots, sensors and the cloud 5th IEEE Intl. Conference on Cloud Networking, 2016.

[53] SAP Leonardo, online at https://www.slideshare.net/PierreErasmus/sap-leonardo-iot-overview

[54] GE Digital's Predix Machine, online at https://www.ge.com/digital/predix

[55] IBM Watson IoT, online at https://www.ibm.com/internet-of-things/partners/ibm-cisco

[56] Amazon AWS Greengrass, online at https://aws.amazon.com/greengrass/

[57] P. Simoens, et al. Internet of Robotic Things: Context-Aware and Personalized Interventions of Assistive Social Robots. IEEE CloudNet, 2016.

[58] G., Fortino, et al. "Middlewares for Smart Objects and Smart Environments: Overview and Comparison". *Internet of Things Based on Smart Objects, Springer, 2014.*

[59] A.G., Thallas, et al; "Relieving Robots from Their Burdens: The Cloud Agent Concept (Short Paper)," *2016 5th IEEE International Conference on Cloud Networking (Cloudnet)*, Pisa, 2016, pp. 188–191.

[60] D, Hunziker, et al; "Rapyuta: The RoboEarth Cloud Engine," *2013 IEEE International Conference on Robotics and Automation*, Karlsruhe, 2013, pp. 438–444.

[61] S., Schmid, A. Bröring, D. Kramer, S. Kaebisch, A. Zappa, M. Lorenz, Y. Wang & L. Gioppo (2017): An Architecture for Interoperable IoT

Ecosystems. 2nd International Workshop on Interoperability & Open Source Solutions for the Internet of Things (InterOSS-IoT 2016) at the 6th International Conference on the Internet of Things (IoT 2016), 7. November 2016, Stuttgart, Germany. Springer, LNCS. Volume 10218, pp. 39–55.

[62] A., Bröring, S. Schmid, C.-K. Schindhelm, A. Khelil, S. Kaebisch, D. Kramer, D. Le Phuoc, J. Mitic, D. Anicic, E. Teniente (2017): Enabling IoT Ecosystems through Platform Interoperability. IEEE Software, 34(1), pp. 54–61.

[63] E. Piscini, G. Hyman, and W. Henry, Blockchain: Trust economy, Tech Trends 2017, online at https://dupress.deloitte.com/dup-us-en/focus/tech-trends/2017/blockchain-trust-economy.html

[64] Sanfeliu, Alberto, Norihiro Hagita, and Alessandro Saffiotti. "Network robot systems." Robotics and Autonomous Systems 56.10 (2008): 793–797.

[65] Stroupe, Ashley W., Martin C. Martin, and Tucker Balch. "Distributed sensor fusion for object position estimation by multi-robot systems." Robotics and Automation, 2001. Proceedings 2001 ICRA. IEEE International Conference on. Vol. 2. IEEE, 2001.

[66] Rosencrantz, Matt, Geoffrey Gordon, and Sebastian Thrun. "Decentralized sensor fusion with distributed particle filters." Proceedings of the Nineteenth conference on Uncertainty in Artificial Intelligence. Morgan Kaufmann Publishers Inc., 2002.

[67] LeBlanc, Kevin, and Alessandro Saffiotti. "Multirobot object localization: A fuzzy fusion approach." IEEE Transactions on Systems, Man, and Cybernetics, Part B (Cybernetics) 39, no. 5 (2009): 1259–1276.

[68] Charrow, Benjamin. "Information-theoretic active perception for multi-robot teams." PhD diss., University of Pennsylvania, 2015.

[69] Burgard, Wolfram, Mark Moors, Cyrill Stachniss, and Frank E. Schneider. "Coordinated multi-robot exploration." IEEE Transactions on robotics 21, no. 3 (2005): 376–386.

[70] Durrant-Whyte, Hugh and Tim Bailey. "Simultaneous localization and mapping (SLAM): Parts I." IEEE Robotics & Automation Magazine 13, no. 2 (2006): 99–110.

[71] Nüchter, Andreas, Kai Lingemann, Joachim Hertzberg, and Hartmut Surmann. "6D SLAM—3D mapping outdoor environments." Journal of Field Robotics 24, no. 8–9 (2007): 699–722.

[72] Krajník, Tomáš, Jaime P. Fentanes, Joao M. Santos, and Tom Duckett. "Fremen: Frequency map enhancement for long-term mobile robot

autonomy in changing environments." IEEE Transactions on Robotics (2017).

[73] Galindo, Cipriano, Juan-Antonio Fernández-Madrigal, Javier González, and Alessandro Saffiotti. "Robot task planning using semantic maps." Robotics and autonomous systems 56, no. 11 (2008): 955–966.

[74] P&S Market Research, "Global Personal Robots Market Size, Share, Development, Growth and Demand Forecast to 2022", 2017. Retrieved from https://www.psmarketresearch.com/market-analysis/personal-robot-market

[75] SPARK, the partnership for robotics in Europe, "Robotics 2020 Multi-Annual Roadmap", 2015. Retrieved from https://www.eu-robotics.net/

[76] PC World, "Domestic Robots: High-Tech House Helpers, 2012," Retrieved from http://www.pcworld.com/article/253882/domestic_robots_high_tech_house_helpers.html.

[77] Integrating Robotics as an Interdisciplinary-Educational Tool in Primary Education, Nikleia Eteokleous-Grigoriou and Christodoulos Psomas, SITE 2013.

[78] M., Bonaccorsi, L., Fiorini, F., Cavallo, A., Saffiotti, and P. Dario, (2016). A cloud robotics solution to improve social assistive robots for active and healthy aging. International Journal of Social Robotics, 8(3), 393–408.

[79] Siciliano, Bruno and Oussama Khatib, editors. Springer Handbook of Robotics (2nd edition). Springer International Publishing, 2016.

[80] Luo, Ren C., Chih-Chen Yih, and Kuo Lan Su. "Multisensor fusion and integration: approaches, applications, and future research directions." IEEE Sensors journal 2, no. 2 (2002): 107–119.

[81] S. LaValle, Planning Algorithms, Cambridge University Press, 2006.

[82] Rajan, Kanna and Alessandro Saffiotti. Towards a science of integrated AI and Robotics. Artificial Intelligence 247:1–9, 2017.

[83] Robotics 2020 Multi-Annual Roadmap (MAR), online at https://www.eu-robotics.net/cms/upload/downloads/ppp-documents/Multi-Annual_Road map2020_ICT-24_Rev_B_full.pdf

[84] Yole Développment Report, "Sensors for drones and robots: market opportunities and technology revolution 2016" Report by Yole Développement, February 2016 From Technologies to Market Sensors for Drones & Robots 2016 report sample.

[85] LoRa Alliance, at https://www.lora-alliance.org/

[86] 3GPP – The 3rd Generation Partnership Project at http://www.3gpp.org/

5

STARTS – Why Not Using the Arts for Better Stimulating Internet of Things Innovation

Peter Friess and Ralph Dum

European Commission, Belgium*

"Artists should be incorporated as catalysts for new ways of thinking, not only about art, but about the world we live in, to change the way things are done, made and developed in the world."

Camille Baker, FET-Art Project

Abstract

Ongoing digital transformation is profoundly changing industry, science and technology. Linking technology and artistic practice is today being considered a win-win exchange between European innovation policies and the art world, up to the point that it would be counterproductive to restrict artistic freedom and independence through the current linear and incrementally oriented patterns of invention and production. It is evident that the Internet of Things (IoT) is one of the biggest game changers for modern societies, whereby the IoT deployment is highly challenging when it comes to address all the industrial domains and the needs of the stakeholders and end-users. And here it is precisely where artists and artistic practices can team up with product and process innovators, because their stimulating ideas and out-of-the-box thinking contribute to a successful transfer into the era of a hyper-connected and potentially more sustainable society.

*The views expressed in this article are purely those of the authors and not, in any circumstances, be interpreted as stating an official position of the European Commission.

157

5.1 Introduction

Europe has historically focused its attention in engineering on research, development and standardisation. Today, an increasing number of high tech companies assert that, in addition to knowledge, creativity is central to companies' and society's ability to innovate. For innovation to happen and to be of value for society, the critical skills needed – in addition to scientific and technological skills – are skills such as inventiveness, and capacity to involve all members of a society in the process of innovation.

In this context, the ongoing digital transformation is profoundly changing industry, science and technology. It can be observed that digitisation is indeed uniting science and engineering with design and the arts, that the boundaries between art and engineering are removed, and creativity has become a crucial factor in engineering and innovation in general. Nowadays, the Arts are gaining prominence as catalysts for an efficient conversion of Scientific & Technical knowledge into novel products, services, and processes.

The European Commission has repeatedly pointed to digital transformation of industry, culture and society as a driver for an innovation-focused cross-sectorial exchange. For radical market driven innovation, industrial players are encouraged to think in a more holistic way in terms of services and of technologies. Linking technology and artistic practice is today being considered a win-win exchange between European innovation policies and the art world. Such links will help overcome current linear and incrementally oriented patterns of invention and production. Those synergies will only work if artistic freedom and independence are not restricted which is the main asset form which we can draw inspiration.

It is certain that the future will be different in the way we create, perceive, communicate and earn our living. Although it has been repeatedly said that the Internet of Things is one of the biggest game changers for modern societies, it is only slowly that actors in many fields, be it for example agriculture, urban life, health, transport and environment, explore concrete avenues on how to use the Internet of Things and redesign their way of operation. So far it has not been fully appreciated that pursuing Internet of Things deployment is not more of the same but that it challenges all of us to approach the future in novel ways. And here it is precisely where artists and artistic practices can team up with product and process innovators, because their stimulating ideas and out-of-the-box thinking will contribute to a successful transition into the era of a hyper-connected and potentially more sustainable society.

5.2 The STARTS Initiative

Europe has a rich artistic heritage and diverse art scene, but this asset is currently underused in promoting innovation and wellbeing in Europe. One particular angle of this deficit is the continued divide between artistic practices and technological knowhow. The resulting deficit is perceived as opposing modes of thinking of art and technology and is enacted in a still prevalent reluctance for arts and technology to collaborate on important challenges for our society.

This has serious consequences for the role of technology and art in innovation in Europe. A number of studies show that creativity is key for innovation and can be unlocked by collaboration between art and technology. An enhanced collaboration between art and technology for the Internet of Things would not only stimulate innovation and thereby enhance the competitiveness of Europe on a global scale; it would also help unleash creativity in our society and in European regions (see Figure 5.1).

However, until recently the acceptance of a transversal cooperation between scientists, technologists and artists was little and best practice cases on how to best stimulate this cooperation were rare and not widely known. For this reason, the European Commission contracted a study about "ICT

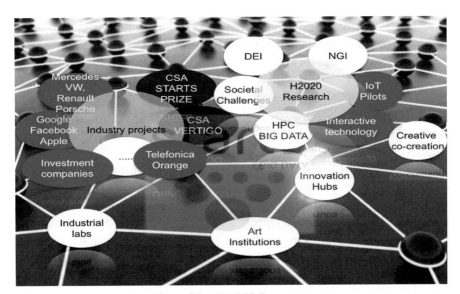

Figure 5.1   STARTS Ecosystem.

Art Connect – Activities linking ICT and Art" [1] in order to better target specific actions. Several barriers were identified: different working cultures between artists and innovators, different timescales, fear of artists to be instrumentalised, legal and contractual issues, non-explicit reference to the Arts in call for proposal texts, financing difficulties, missing openness on both sides, and lack of training and meta-competences.

As a consequence, the European Commission launched in 2016 the S+T+ARTS=STARTS [2] initiative – Innovation at the nexus of Science, Technology, and the Arts. Its objective is to provide seed funding for enhancing the interaction of H2020 projects with the art world and to promote inclusion of artists in innovation projects funded in H2020 and beyond. In a next step, the initiative should provide case studies – concrete projects where the Arts catalyse the novel application of technology in fields like Internet of Things, or in Social Media, where the influence of artists on novel uses and applications is strong.

5.2.1 STARTS Prize

The STARTS prize is awarded every year to projects at the cutting edge of creative and cultural engagement. In order to diversify the prize two categories are awarded:

- **Artistic Exploration**, where by appropriating technology in their artistic exploration artists help open new pathways for technology,
- **Innovative Collaboration**, where collaborations between artists and engineers are honoured that contribute to innovative product and services development in the context of industrial or societal innovation.

The STARTS prize is currently organised on behalf of the European Commission by Ars Electronica in collaboration with the Centre of Fine Arts in Brussels and Waag Society (see Figure 5.2).

Winners in 2016 were in the first category Iris van Herpen for the use of magnetic force fields as a design tool for clothes and shoes. For the second category the prize was awarded to Ottobock, a medical technology company, and the Berlin Weissenberg Art School for innovation in the design of prostheses ('artificial skins and bones'). For 2017, the awards are attributed to the investigation into the constructive principle of the physical phenomena of jamming for construction, and to the exploration of the concept of 'post-humanity music'.

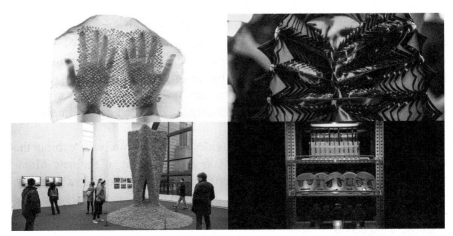

Figure 5.2 STARTS Prizes 2016 and 2017.

5.2.2 VERTIGO Coordination and Support Action

The VERTIGO project is supported by the H2020 Program of the European Commission, and its purpose is to support and coordinate synergies between the art world and the world of engineering and industry at the European level (see Figure 5.3).

Figure 5.3 VERTIGO artistic residencies.

VERTIGO will launch a program for STARTS residencies that will fund artists willing to work in technology environments for a duration of a couple of weeks. This will be done through several calls for proposals. Winners of STARTS residencies will be selected by an international jury. At least 45 residencies will aim at producing original artworks featuring innovative use-cases of the developed technologies in many areas including the Internet of Things.

A yearly public event in Paris provides the opportunity to exhibit the results of these collaborations. It will take place as part of the new Muta-tions/Creation platform initiated at Centre Pompidou, gathering exhibitions, performances and symposia, and dedicated to exposing and questioning the current challenges of contemporary arts in relation to their technological and scientific ecosystem.

5.2.3 Internet of Things European Large-Scale Pilots Programme

The Internet of Things European Large-Scale Pilots Programme includes innovation consortia that are collaborating to foster the deployment of solu-tions in Europe through integration of advanced Internet of Things technolo-gies across the value chain, demonstration of multiple applications at scale and in a usage context, and as close as possible to operational conditions. At present the Programme addresses the fields of Smart Agriculture, Assisted Living, Wearables, Smart Cities and Autonomous Vehicles (see Figure 5.4).

The Pilots are accompanied by a dedicated supporting task as part of the project CREATE-IOT to develop a methodology for integrating Internet of Things and the arts, and to foster innovation in a heterogeneous Internet of Things ecosystem through the STARTS approach. The goal is to inject activ-ities involving arts and artists that will lead to a more successful deployment of the Large-Scale Pilots results to the market.

Figure 5.4   European Large-Scale Pilots Programme.

The activities include analysis of barriers to Internet of Things adoption and to follow-up user acceptance for improvement. They also intent to question the meaning and objectives of the pilots for a more sustainable implementation. Furthermore, the two Large-Scale Pilots for Wearables and Smart Cities are outfitted with a dedicated component on how to structurally integrate artists into the project realisation.

5.2.4 STARTS Lighthouse Pilots

The ultimate goal is to support art-driven innovation in European research and to create a STARTS ecosystem involving all stakeholders of 21st century innovation, industry, engineering, end-users, and in particular artists. To this end, the European Commission will finance lighthouse pilots addressing concrete industrial/societal problems, exploring new pathways and modes of thinking inspired by artistic practices, and developing art-inspired solutions in two chosen areas which can be situated around the Internet of Things (see Figure 5.5).

The lighthouse pilots will address in a first phase two areas:

- Digital objects and media for creation of smart environments for homes, mobility or urban spaces, putting art-driven design and development of

Figure 5.5   Future STARTS lighthouse pilots – catalysts of human-centred innovation [3].

services and products in the centre of radically different human-centred smart environments,

- Future of small scale production and co-creation value chains for reviving the urban social, ecological and economic spaces.

5.3 Internet of Things and the Arts

A successful implementation and non-technical deployment of the Internet of Things considers both technical and non-technical challenges. Although this appears trivial, the reality is dominated by a technology push, driven by efficiency justification and competition, and strategies for revenue increase. What is left behind, are pivotal reflections on the proper purpose and use of the Internet of Things, about sustainability (e.g. "less and differently than more of the same"), security and trust, about creativity where more and more ICT causes a suffocation of free thinking and human behaviour, not to forgot ethical questions about permanent data collection, implants, robots and artificial intelligence. If those questions are not properly addressed any implementation of the Internet of Things will miss its full potential.

Generally, the Arts can contribute on three levels towards better and sustainable evolution and innovation (see also Figure 5.6):

- **Catalysing function for innovation**: e.g. design of wearables, objects and smart buildings
- **Reinforcement of end-user and social engagement**: e.g. community-based development, digital empathy, green attitude
- **Critical attitude for pinpointing weaknesses of a system and its implementation**: e.g. critical documentary, creation of social media movements, system testing and hacking

Figure 5.6 Different levels of artistic collaboration of the Internet of Things [4–6].

Now where in the context of the Internet of Things the Arts could beneficially intervene? As possibilities are numerous, the following list can only be considered as a starting point and needs to be extended.

- **User experience and Interface design**: through the Internet of Things the future user experience is potentially twofold: technical devices and machines are disappearing, as e.g. environments become smart, but we also will deal with more of existing, augmented or new objects. In both cases, new and disruptive user experiences are to be designed where the artistic perspective could be highly beneficial.
- **Solution design**: here the objectives and functions are decisive for successful Internet of Things application take-up. A creative and critical artistic perspective could help for radical innovation, but also to get out of a technology-only driven approach.
- **System design**: in order to satisfy specified requirements, artists could help improving the process of defining architectures, modules, interfaces, and data bases from various angles, such as new creative perspectives, asking un-orthodox questions and making critical remarks e.g. concerning the extent of use of data.
- **Testing and Improvement**: testing is crucial for pushing the Internet of Things system towards adoption. Often on testing and subsequent improvements are neglected or even being considered as annoyance. In this context artists could e.g. take a mediating role both for a feedback process and supporting a common understanding.
- **Communication**: the importance of communication is often underestimated for a project success. But instead of wondering about professional support, artists could create/invent unconventional means and images for better communication.
- **Ethical/societal questions**: particularly for the Internet of Things and in light of a highly transversal deployment these interrogations play a crucial role. Artists can pinpoint and help to thematise them as they might be pushed back and repressed by the project team.

It should not be forgotten that the cooperation between artists and innovators is not happening per se, even as in theory positive results can be expected. Efforts need to be made on how those different working cultures can be brought together and synchronised. It is important to address the different timescales between engineers and artists, rapid remuneration, team-building, training and advancing through iterative steps and pilots.

5.4 Conclusion

The STARTS initiative supports the artistic practice contributing to innovation of information and communication technologies. It emerged out of the necessity to facilitate the cooperation between artists and innovators at the nexus of Science, Technology, and the Art. At present the catalytic aspect of artistic practices in innovation projects has been proven, although hesitations remain. The creation and extension of the STARTS Ecosystem for increasing awareness and the number of collaborations remains at the heart of the European Commission. The Internet of Things is potentially one of the most promising areas of cooperation, which at the same time due to its societal importance has a huge demand for artistic intervention and therefore presents a perfect match.

References

[1] https://ec.europa.eu/digital-single-market/en/news/innovation-about-starts-when-ict-and-art-connect

[2] http://www.starts.eu/

[3] https://medium.com/iotforall/artificial-intelligence-machine-learning-and-deep-learning-169a4a136f62

[4] https://www.ariasystems.com/blog/monetizing-wearables-data/

[5] http://www.gtreview.com/news/global/the-internet-of-things-and-the-future-of-logistics/

[6] https://iot-for-all.com/designing-the-internet-of-things/

6

IoT Standards Landscape – State of the Art Analysis and Evolution

Emmanuel Darmois[1], Laura Daniele[2], Patrick Guillemin[3], Juergen Heiles[4], Philippe Moretto[5] and Arthur Van der Wees[6]

[1]CommLedge, France
[2]TNO, The Netherlands
[3]ETSI, France
[4]Siemens AG, Germany
[5]Sat4m2m, Germany
[6]Arthur's Legal B.V., The Netherlands

6.1 Introduction

The Internet of Things (IoT) is now more than an emerging technology, and the IoT community has started to develop ambitious solutions and to deploy large and complex IoT systems. However, this new challenge for IoT will be met only if the IoT community develops a culture of openness regarding interoperability, support of a large variety of applications departing from existing silos, and the generation of healthy ecosystems.

The role of standards is now well recognized as one of the key enablers to this open approach. There are already a number of existing standards for those who develop IoT systems. They allow to address many of the requirements of IoT systems in a large spectrum of solutions (ranging from consumer to industrial) for a large number of domains, as various as cities, e-health, framing, transportation, etc.

The objective of this chapter is to make an overview of the current state-of-the-art in standardisation, in particular regarding the new approaches that are currently addressed by standards organisations and that will rapidly enlarge the scope of current standards. On the other hand, given the complexity of the IoT landscape, some elements are still missing and will need to be addressed as well: another objective of this paper is to provide an overview of those gaps and how they may be resolved in the near future.

6.2 IoT Standardisation in the Consumer, Business and Industrial Space

The IoT community has recognized long ago the importance of IoT standardisation and started to work in many directions, adapting general purpose standards to the IoT context or developing new IoT specific standards. There is now a large number of standards that can be used by those who want to develop and deploy IoT systems. This section will address the current state-of-the-art, evaluate the number of available standards and suggest ways to classify them.

When a large number of standards exist in a given domain, there is a risk of duplication, fragmentation, competition between standards organisations. In its 2016 communication on "ICT Standardisation Priorities for the Digital Single Market" [1], the European Commission notes that: "However, the IoT landscape is currently fragmented because there are so many proprietary or semi-closed solutions alongside a plethora of existing standards. This can limit innovations that span several application areas".

This is a major challenge for IoT standardisation: because IoT is a large domain, spanning across a variety of sectors (e.g., food, health, industry, transportation, etc.), many standards potentially apply that have been developed within application silos, and the risk of fragmentation exists. The European Commission also outlines an essential way-forward [1]: "Large-scale implementation and validation of cross-cutting solutions and standards is now the key to interoperability, reliability and security in the EU and globally".

Two complementary dimensions (outlined in the next subsections) are taken into account by the IoT standardisation community:

- Expansion of the reach of "horizontal layers" standards versus "vertical domains"-specific standards;
- Specialisation of general purpose standards for application to more complex and demanding domains. This is in particular the case with the convergence of IT (Information Technology) and OT (Operational Technology) in the industrial domain.

6.2.1 Standardisation in Horizontal Layers and Vertical Domains

The IoT landscape developed by the AIOTI Work Group 3 on Standardisation has used the distinction between the horizontal and vertical domains for the classification of the organisations that are active in IoT standardisation.

The classification of IoT standardisation organisations is done along two dimensions:

- Vertical domains (or "verticals") that represent 8 sectors where IoT systems are developed and deployed;
- An "horizontal" layer that groups standards that span across vertical domains, in particular regarding telecommunications.

In order to give an indication of the relative importance of "horizontal" versus "vertical" standards, the ETSI Specialist Task Force (STF) 505 report on the IoT Landscape [4] has identified 329 standards that apply to IoT systems. Those standards have been further classified in:

- 150 "Horizontal" standards, mostly addressing communication and connectivity, integration/interoperability and IoT architecture.
- 179 "Vertical" standards, mostly identified in the Smart Mobility, Smart Living and Smart Manufacturing domains.

One important way to ensure that "interoperability, reliability and security" aspects (outlined above as key by the EC report [1]) are handled more efficiently is to make sure that "horizontal" standards are chosen over "vertical"

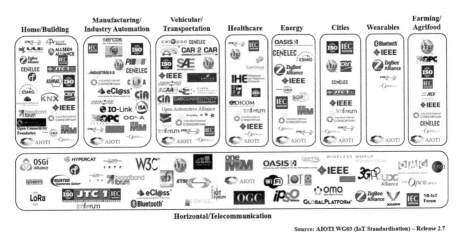

Figure 6.1 IoT SDOs and Alliances Landscape.

ones whenever possible. An "horizontal" standard is likely to be developed to serve general-purpose requirements and better address interoperability.

In addition to the IoT Standardisation landscape, the AIOTI has developed the High-Level Architecture (HLA) that defines three layers (as depicted in Figure 6.2 below) and provides more complete ways to characterise and classify the applicable standards:

- The Application layer contains the communications and interface methods used in process-to-process communications
- The IoT layer groups IoT specific functions, such as data storage and sharing, and exposes those to the application layer via Application Programming Interfaces (APIs). The IoT Layer makes use of the Network layer's services.
- The Network layer services can be grouped into data plane services, providing short and long range connectivity and data forwarding between entities, and control plane services such as location, device triggering, QoS or determinism.

The HLA supports a more fine-grain classification of "horizontal" standards that are in general addressing only one of the HLA layers, thus offering a clearer scope for interoperability.

The IoT Standardisation Landscape in Figure 6.1 clearly shows that the "horizontal" standards are developed by organisations (SDOs/SSOs) that deal with IT technology solutions rather than by those operating in "vertical" domains. The potential of "horizontal" standards (common standards across vertical domains) will only materialize if the development of IoT standards in vertical domains is making effective use of those standards rather than reinventing similar but not compliant ones. On the other hand, collaboration and cooperation between the SDOs/SSOs involved in "horizontal" standards must be encouraged.

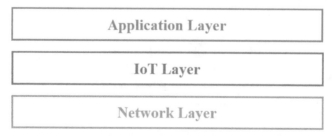

Figure 6.2   AIOTI three layers' functional model.

6.2.2 Standards Addressing the Convergence of IT and OT

IoT standardisation has been a huge effort of many organisations (vendors and manufacturers, service providers, brokers, etc.). Two main kind of activities have taken place for the development of "horizontal layer" standards as well as "vertical domains" standards.

On the one hand, the "horizontal layer" standards have been mostly developed by the Information and Communication Technologies (ICT) industry with a particular focus on Information Technologies (IT), in particular those associated to communications and to new deployment models such as the cloud.

On the other hand, the "vertical" domains have started to address the requirements of IoT with the goal to expand the reach of the existing domain-specific standards. The resulting IoT standards have been coming from the massive incorporation of IT technologies, whether by adapting the existing standards or by adopting ICT standards.

IoT standardisation has addressed growing levels of complexity depending on the nature of the IoT systems concerned. Many of the ICT standards have been rapidly adopted in the Consumer space, be it for communications, security or semantic interoperability (see the example of SAREF addressed in Section 6.3.2). The requirements of IoT systems in the Business space requires an additional degree of complexity in order to be able to deal with complex data models, strict security, privacy or large scale deployments.

A new challenge for IoT standardisation is regarding industrial IoT systems (see [5]). The challenge is to massively integrate new technologies such as IoT or Cloud Computing in order to provide much more flexibility, adaptability, security and reliability. This will require achieving a transition from the current model to the "Cyber-Physical Production System" (CPPS) approach.

The current model is based on the "Manufacturing pyramid" approach, with hierarchically separated layers, where the interactions between the bottom layer of IoT devices and the upper layer of the production system are complex with supporting data models often too much specialized.

In Cyber-Physical Production Systems (CPPS), the field level (e.g., the factory, the robots, the sensors) will be connected to a wide range of applications and services – using of vast quantities of data available to plan, monitor, re-tool and maintain, etc. – together with being ensured a higher level of reliability, as well as trust and security from a redefined security architecture.

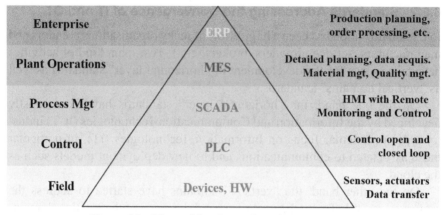

Figure 6.3   The traditional manufacturing pyramid [6].

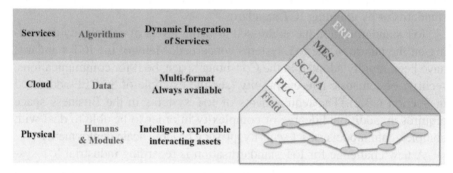

Figure 6.4   Cyber Physical Production Systems [6].

The current model is well covered in standardization [4]. The transition towards CPPS has started and will require a major leap forward in the integration of IT (Information Technology) with OT (Operational Technology: "industrial control system and networks, hardware and software that detects or causes a change through the direct monitoring and/or control of devices, processes, and events in the enterprise [5]).

6.3 New Trends in IoT Standardization

IoT standardisation is a large effort with a lot of parallel undertakings. In order to address the challenges outlined above, some topics of special interest are emerging. This section intends to address some of them, in particular:

- Identification and addressing. With systems growing in size and complexity, the need to ensure that all devices are well identified and properly addressable becomes a key requirement.
- Semantic Interoperability. Many of the current systems are based on static data models. The promise of semantic interoperability is to ensure much more dynamic data models. Its challenge is to make sure that the approach is scalable and can be used in real-life IoT systems deployments.
- Security and Privacy. Though both topics are different in scope and in the solutions developed, they share a common characteristic: security and privacy are make-or-break for large IoT systems deployments and user adoption.

6.3.1 Identification and Addressing in IoT

In any system of interacting components, identification of these components is needed in order to ensure the correct composition and operation of the system. This applies to the assembly and commissioning of the systems, and is also relevant for system operations, especially in case of flexible and dynamic interactions between system components. In addition, identification of other entities like data types, properties, or capabilities is needed; however, that is related to semantics expressions and ontologies for such entities and not to dedicated identifiers.

IoT systems provide interaction between users and things. In order to achieve this, device components (sensors and actuators), service components, communication components, and other computing components are needed, as shown in Figure 6.5. The virtual entity plays a special role in IoT as it provides the virtual representation of things in the cyber world; it is closely linked to the thing for which identifiers are essential.

In general, an identifier is a pattern to uniquely identify a single entity (instance identifier) or a class of entities (type identifier) within a specific context. Figure 6.5 shows some examples of identifiers for IoT.

Things are at the centre of IoT and unique identification of Things is a prerequisite for IoT systems. Kevin Ashton who coined the term "Internet of Things" in 1999 linked the term with identification, specifically Radio Frequency Identification RFID. RFID is one means of identification, but many more exist, given that a *thing* could be any kind of object:

- Goods along their lifecycle from production to delivery, usage, maintenance until end of life

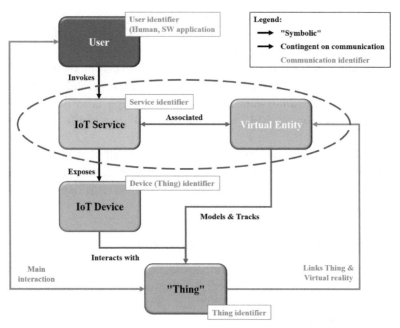

Figure 6.5   Identifiers examples in the IoT Domain Model (according to AIOTI WG3 High Level Architecture).

- Weather conditions in a certain area
- Traffic flow at an intersection
- Vehicles and containers for tracking purposes
- Animals and field yields for smart farming applications
- Humans in case of health and fitness applications
- Digital objects like e-books, music and video files or software

Some of the things are directly connected to a communication network while others are only indirectly accessed via sensors and actuators. Identification can be based on inherent patterns of the thing itself like face recognition, fingerprints or iris scans. In most cases a specific pattern will be added to the thing for identification by technical means like printed or engraved serial numbers, bar codes, RFIDs or a pattern stored in the memory of devices. As identification applies to systems in general, many identification means already exist and are often standardized as domain specific solutions. Users may prefer different identification schemes. A property management company identifies things according to the building location, floor and room number while the producer of the thing uses its own serial number scheme.

Furthermore, the identifier might be unique within its current usage context, but new applications may result in conflicts if the same identifier is used in other systems. This raises issues of interoperability, uniqueness and linkage between identifiers which need further elaboration.

Users that interact with the things could be humans or software applications. Identification of the users is needed, especially if access is limited and/or tracked. From a security point of view, authentication requires a second step to validate the claims asserted by the identity. Privacy concerns must also be considered.

In the case of communication networks, the source and destination of the communication relationships must be identified. Here the identifiers are bound to the specific communication technology and defined as part of the standardization of the technology. IP networks use IPv4 and IPv6 addresses, Ethernet and WLAN use MAC addresses and fixed and mobile phone networks use phone numbers. Communication identifiers may not be a good fit as identifiers for *things* as the communication address of a thing may change (e.g. the communication interface or network topology may change if a different communication service provider is selected). Furthermore, some things don't have communication interfaces whilst others may have more than one (e.g. for redundancy reasons).

The AIOTI WG03 IoT Identifier task force will evaluate and classify identification needs and related requirements for IoT. As a part of this task, existing identification standards and ongoing standardization work will be examined and the applicability for the different identification needs will be elaborated. The task force has performed an open survey with various standardization bodies, research activities, industry associations, companies and individuals concerning the above topics. The results of the survey will contribute to a white paper on IoT identification needs, requirements and standards. Security, privacy and interoperability issues will also be considered and standardization gaps will be analysed. A first version of the White Paper is expected for end of 2017.

6.3.2 The Challenge of Semantic Interoperability

The Internet of Things (IoT) is happening all around us, not the least within the home and the (smart) city. IoT devices such as tablets, thermostats, energy meters, lamp systems, home automation, washing machines, motion sensors and personal health devices, are nowadays easily bought by consumers in the (web-) shops and are connected via the Internet to third-party service provider's systems. In order to provide consumers of the smart home and citizens of the smart city a true and seamless IoT experience, however, the multitude of IoT devices, systems and services need to be interoperable, i.e., able to exchange information with one another and use the information so exchanged[1]. According to a recent McKinsey report, interoperability is essential to unlock 40% of the $11 trillion potential value of the IoT [7].

In the past years, standards organizations, industry alliances and consortia have focused on technical interoperability[2], which covers basic connectivity, network interoperability and syntactic interoperability, so that devices can exchange messages based on a common syntax [8]. The industry tacitly assumed that communicating devices would have a common understanding of the meaning of exchanged messages, which is true within the same ecosystem and for the same domain, as long as the design engineers and devices of the different companies speak the same technical language. However, the integration of systems for different domains with different ecosystems in the background requires a coordinated approach of the involved partners in order to agree on the meaning of the information contained in the message data structures, regardless of which communication protocol they are based on. In other words, interoperability at the technical communication level is no longer sufficient and there is a need for **semantic interoperability**.

To achieve semantic interoperability, all manufacturers involved must refer to a (set of) commonly agreed information exchange reference model(s), which not only contains the syntax, but also the meaning (semantics) of the concepts being used. By creating interoperability on the semantic level, it becomes possible to translate information to, from and between devices, thereby making it possible to control them in a standardized way. The need to address the semantics of standards has been acknowledged as an important action in the upcoming IoT standardization activities towards an interoperable and scalable solution across a global IoT ecosystem [9].

A powerful way to represent such common models and support the current standardization activities in the IoT is the use of standardized common vocabularies, or ontologies, which can formally represent the semantics

---

[1]Interoperability as defined by the European Information & Communications Technology Industry Association (EICTA), now called DIGITALEUROPE, also adopted by CENELEC TS 50560.

[2]Technical interoperability as defined by the GridWise Interoperability Framework, also adopted by AIOTI WG03.

of concepts exchanged by different devices and ecosystems. The Smart Appliances REFerence ontology (SAREF)[3] serves as a successful example[4].

In 2013, the European Commission launched a standardization initiative in collaboration with ETSI to create a shared semantic model of consensus to enable the missing interoperability among smart appliances. The focus of the initiative was to optimize energy management in smart buildings, as more than 40% of the total energy consumption in the European Union comes from the residential and tertiary sector of which a major part are residential houses (therefore appliances, that are inherently present in the buildings' ecosystem can be considered the main culprits of this high energy consumption). TNO was invited to lead this initiative and carry out the work in close collaboration with smart appliances manufacturers and ETSI (Jan 2014–Apr 2015). The resulting semantic model – SAREF – was standardized by ETSI in November 2015 (TS 103 264) [10]. As confirmed in [9], SAREF is a first ontology standard in the IoT ecosystem, and sets a template and a base for the development of similar standards for the other verticals to unlock the full potential of IoT.

SAREF is to be considered as an addition to existing communication protocols to enable the translation of information coming from existing (and future) protocols to and from all other protocols that are referenced to SAREF. As an example, a SAREF-enriched home gateway associates devices in a home with each other and with different service providers. The role of the network operator is to enrich their home gateways with a SAREF-based execution environment, as well as guarding the privacy of customers (as the home gateway has become an omniscient device). A study recently launched by the European Commission (SMART 2016/0082)[5] will demonstrate an implementation for interoperability for Demand Side Flexibility that uses SAREF and its extension SAREF for Energy to enrich a home gateway with additional semantics to be embedded in e.g. oneM2M resources for transportation at the underlying technical level.

Since its first release in 2015, SAREF has gradually grown into a modular network of standardized semantic models[6] that continues to evolve systematically within the SmartM2M TC in ETSI [11]. The first 3 extensions that have been standardised are SAREF for Energy [12], SAREF for Environment [13]

---

[3]http://w3id.org/saref

[4]https://ec.europa.eu/digital-agenda/en/blog/new-standard-smart-appliances-smart-home

[5]https://ec.europa.eu/digital-single-market/en/news/study-ensuring-interoperability-enabling-demand-side-flexibility-smart-20160082

[6]www.ec.europa.eu/digital-single-market/news-redirect/57284

and SAREF for Buildings [14] and a multitude of other domains such as Smart Cities, Smart AgriFood, Smart Industry and Manufacturing, Automotive, eHealth/Ageing-well and Wearables, are on the roadmap turning SAREF into "Smart Anything REFerence ontology", which enables better integration of semantic data from various vertical domains in the IoT.

Another relevant standardization initiative on semantic interoperability for the IoT is the Web of Things[7] (WoT) promoted by the World Wide Web Consortium (W3C) as a way to combine the Internet of Things with the Web of data in order to counter the fragmentation of the IoT through standard complementary building blocks (e.g., metadata and APIs) that enable easy integration across IoT platforms and application domains. In February 2017, W3C launched a new Web of Things Working Group[8] – as an evolution of the Interest Group that has been active in the past years – to develop initial standards for the Web of Things, tasked with the goal to counter fragmentation, reduce the costs of development, lessen the risks to both investors and customers, and encourage exponential growth in the market for IoT devices and services. The proposition of the W3C Web of Things Working Group is to describe things in the IoT in terms of actions, properties, events and metadata (as those are common aspects shared by the vast majority of connected devices), and independent of their underlying IT platforms, trying to complement the work that different organizations are doing, developing cross-domain Linked Data vocabularies, serialization formats, and APIs. An overview for implementers of the W3C WoT building blocks is published in [15], which provides an unofficial draft of the current WoT practices in a single location. While [15] is not a technical specification, it aims at helping implementers to get an overview of the WoT building blocks and includes reports from past PlugFests and follow-up discussions, which explain the rationale behind the WoT current practices.

6.3.3 Addressing Security and Privacy in IoT

Trust, Data Protection and Resilience in IoT as Key Components
Technology changes the world at a fast pace and massive scale. Digital technology makes innovation possible in our society and economy. Cloud computing, data analytics, AI and Internet of Things (IoT) will expedite this pace by hyper-connecting people, organizations and data with billions

---

[7]http://www.w3.org/WoT/
[8]https://www.w3.org/WoT/WG/

of objects. In such diverse physical-cyber, cyber respectively cyber-physical ecosystems, it remains to be seen how demand side, supply side, policy makers, law enforcement, authorities as well as end-users and other stakeholders are going to understand, build, deploy and use IoT ecosystems and its related products, systems and services. Trust remains one of the main challenges of any technology, and given (a) the data-centric nature of IoT products, systems and services, (b) the fact that such data is to a large extent highly sensitive, personal or otherwise valuable to individuals, companies and organisations, and (c) the fact that digital technologies are nowadays a need to have, and individuals, companies and organisations fully rely – and need to be able to fully rely – on these, security and resilience as well as privacy and data protection are key components to trustworthiness. Therewith, Trust, Data Protection and Resilience can be seen as key enablers to build, shape, monitor and optimize trust and with that successful engagement by and with all relevant stakeholders.

Main Categories in Human-Centric IoT

There are several ways to segment and therewith make more transparent and understandable any technology, including IoT. Without such segmentation and classification, it is quite difficult to get on the same page and ensure that interdisciplinary collaboration with stakeholders with multiple backgrounds and expertise on topics such as design, engineering, architectures, governance, risk management, impact-based measures, user-adoption, standards and other policies are focussed and fruitful, especially when addressing trust, security, privacy and (personal) data protection. This obviously, while taking into account, that at the end, these segments need to be and are integrated, hyper-connected and interoperable as detailed in the other paragraphs of this Chapter. A way of segmenting IoT in four main categories is the following:

A. **Data** (including data, information and knowledge)
B. **Algorithms** (including as code, software and services)
C. **Machines** (including devices and hardware systems)
D. **Computing** (including high performance computing, systems and communication of any kind)

In Human-centric IoT, the above main categories are of course each complemented with that human-centric dimension, including for instance human rights, consumer rights, user-interfaces, identity, authentication and the Human Factor, even when there is a purely M2M layer or subdomain involved.

Symbiosis between Security and Data Protection

There is no data protection without security. This goes for both personal data as well as any non-personal data. For instance, personal data protection and privacy is as much about security as it is about data management. Through IoT products, systems and services, organizations create, collect, process, derive, archive and (ideally and to the extent permitted) delete large amounts of data. As part of this lifecycle, digital data is also transmitted, exchanged and otherwise processed around the world, any time, (almost) any place. In short: data likes to travel. Therefore, information security nowadays is not about data ownership but about data control, access, use and digital rights management. Article 29 Working Party, the pan-European body of all data protection authorities ('DPAs') actually have recently stated this as well[9]. With appropriate and dynamic technical and organisational security measures in place it is possible to achieve a dynamic yet appropriate level of personal data protection. In other words, security is a necessary prerequisite for privacy and (personal) data protection. As a consequence, both security and privacy provide essential building blocks for trustworthiness in digital technologies.

Conflict between Human-centric, Data-centric and Process-oriented

More and more organizations are picking up speed to explore how to benefit from digital technology, while mitigating associated risk. From an information security perspective, for more than a decade organisations (whether provider or customer) have taken steps and implemented organizational and technical measures in order to seek and obtain compliance and assurance regarding various international information security standards, such as the ISO 27000 series, SSAE 16 SOC series. However, with the user-centric General Data Protection Regulation (GDPR), just being compliant to those information security standards will not be enough from a GDPR perspective. Any organisation active within the European Union must now apply state of the art security measures (both technical and organizational) where (i) the related cost of implementation, (ii) the purposes of personal data processing

---

[9]BEREC Workshop on Enabling the Internet of Things of 1 February 2017: http://berec.europa.eu/eng/events/berec_events_2017/151-berec-workshop-on-enabling-the-internet-of-things

and (iii) the impact on the rights and freedoms of the data subject (also good, bad and worst case scenarios) need to be taken into account, whether one is either a data controller, co-controller, processor or co-processor. We call this the appropriate dynamic accountability (ADA) formula:

State of the art security – Costs – Purposes + Impact

It is important to note that other than where current information security standards aim at 'achieving continual improvement', the GDPR aims to ensure up-to-date levels of protection by requiring the levels of data protection and security to continuously meet the ADA formula.

An organisation will not be compliant with EU rules unless it follows these user-centric and impact-based requirements. Hence, we are now in an era that where standards traditionally focus on technology-centric processes and controls, new regulation such as the GDPR – soon to be followed by the upcoming ePrivacy Regulation – is user-centric, data-centric and impact-driven. This a new phenomenon and will need to be assessed, addressed and implemented, as the GDPR is a mandatory regulation and one would want to avoid those hefty penalties, which for large enterprise can amount to several billions of Euros. Having analyzed the state of play of international information security standards and its frameworks, we can safely conclude that GDPR raises the bar for personal data protection and related security by introducing user-centric, specific data-centric and impact-based requirements as opposed to process- and technology-oriented frameworks of standards. Being compliant in the traditional way where compliance refers to linear and binary compliancy and assurance is not good enough anymore. Technology has become a highly-regulated domain in itself. The good news is that, once an organization does have those dynamic and appropriate technical and organizational measures in place, it will significantly increase trustworthiness towards customers, users, authorities and other stakeholders, and demonstrates next generation readiness.

Principle-Based Security and Privacy

Combining the vast domain of cybersecurity, security and safety, with the even vaster domain of IoT is a necessity, yet quite complex and difficult to grasp and comprehend.

In order to come to workable and actionable frameworks and models to address the pre-requisite trust components of Security and Privacy in IoT, come to the mandatory level of appropriate accountability (as for instance set forth in the GDPR) and enable organisations in any sector, including public and private, to assess which technical and organisational security measures it needs to consider and implement, various organisations have set up committees, taskforces and workshops. In 2016 and first part of 2017, this has resulted in about 30 papers that describe such recommendations, frameworks and other guidelines on state of the art level in Security in IoT.

The IoT Unit of the European Commission, together with relevant stakeholders including AIOTI and key IoT industrial, demand side and policy players have organised two workshops the past year, including in June 2016 the AIOTI Workshop on Security & Privacy in IoT[10] [17] and the European Commission's Workshop on Security & Privacy in IoT of 13 January 2017 [18], resulting in recommendations, principles and requirements as set forth in its respective reports in order to enable and facilitate the increase of security, privacy, identify minimum baseline principles and requirements for any IoT product, service or system, and therewith trust in human-centric IoT.

One structures and analyse these in the perspective of the following layers and dimensions, where dimensions may be relevant in one, more or all layers:

LAYERS	DIMENSIONS
1. Service	A. User/Human Factor
2. Software/Application	B. Data
3. Hardware	C. Authentication
4. Infrastructure/Network	

The two reports [17] and [18] result in a structured set of about 50 principles and requirements, and also makes visible where appropriate maturity of those principles have been reached, and where possible gaps and points of attention can be identified and how to address those.

A sample of the structured overview, in this case of the hardware layer, of the so-called State of the Art (SOTA) Layered Plotting Methodology is set forth below. The full version of the total overview hereof can be found at www.arthurslegal.com/IoT and www.aioti.eu.

---

[10]AIOTI Workshop on Security and Privacy in IoT of 16 June 2016: https://ec.europa. eu/digital-single-market/en/news/aioti-workshop-security-and-privacy-etsi-security-week

HARDWARE

- **Security principles:**
  - *High-level baseline:* High level baseline should be applied when safety is at stake or critical infrastructure or national safety can be materially impacted.
  - *Separate safety and security:* Manufacturers have to implement and validate safety principles, separately from security principles.
  - *Security rationale:* Manufacturers should be required to provide explanation of implemented security measures related to expected security risks from any designer of IoT device, auditable by independent third party.
  - *Security evaluation:* Manufacturers should specify precisely capabilities of device of a particular type. This could help to manage liability and evolutivity on system level.
  - *Security levels:* The industry should make use of the security scale 0 – 4 fit to the market understanding.
  - *Sustainability:* Manufacturers should ensure that connected devices as well as any IoT component as defined above are durable and maintained as per its purpose, context and respective life cycle.
  - *Assurance:* Component and system suppliers need to be prepared for security monitoring and system maintenance over the entire life cycle and need to provide end of life guarantees for vulnerabilities notifications, updates, patches and support.
- **Certification and Labelling:**
  - *Certification:* Device manufacturers should test devices and make use of existing, proven certifications recognized as state-of-the-art based on assessed risk level. Additional introduction of a classification system to certify devices for use in particular use case scenarios depending on the level of risk should be encouraged.
  - *Trusted IoT label:* Labels such as the 'Energy efficiency label' of appliances should give a baseline requirement of protection based on the level of assurances and robustness, and should be used to classify individual IoT devices.
- **Secure Performance and Functionality:**
  - *Defined functions:* Manufacturers should ensure that IoT devices are only able to perform documented functions, particular for the device/service.
  - *Secure interface points:* Manufacturers should identify and secure interface points also to reduce the risk of security breach.

6.4 Gaps in IoT Standardisation

Despite a large number of available standards on which to build IoT systems, the development of large-scale interoperable solutions may not fully guaranteed, when some elements in the IoT standards landscape are missing. Such elements, commonly referred to as "gaps", are subject of a number of analysis that aim at identifying them with the intent to ensure that their resolution can be handled by the IoT community, in particular the standardisation community.

6.4.1 Identifying IoT Standards Gaps

Though the gaps related to missing technologies are the most commonly thought of, several categories of gaps can be identified and need to be equally addressed. In the work of ETSI STF 505 [15], three categories of gaps have been addressed:

- Technology gaps with examples such as communications paradigms, data models or ontologies, or software availability.
- Societal gaps with examples such as privacy, energy consumption, or ease of use.
- Business gaps with examples such as silo-ed applications, incomplete value chains, or missing investment.

The perceived criticality of the gaps may be different depending on the role of an actor in standardisation. The Table 6.1 below is listing some of the major gaps identified in [15]. In addition to their nature and type, it also provides a

Table 6.1   Some standards gaps and their perceived criticality

Nature of the Gap	Type	Criticality
Competing communications and networking technologies	Technical	Medium
Easy standard translation mechanisms for data interoperability	Technical	Med
Standards to interpret the sensor data in an identical manner across heterogeneous platforms	Technical	High
APIs to support application portability among devices/terminals	Technical	Medium
Fragmentation due to competitive platforms	Business	Medium
Tools to enable ease of installation, configuration, maintenance, operation of devices, technologies, and platforms	Technical	High
Easy accessibility and usage to a large non-technical public	Societal	High
Standardized methods to distribute software components to devices across a network	Technical	Medium
Unified model/tools for deployment and management of large scale distributed networks of devices	Technical	Medium
Global reference for unique and secured naming mechanisms	Technical	Medium
Multiplicity of IoT HLAs, platforms and discovery mechanisms	Technical	Medium
Certification mechanisms defining "classes of devices"?	Technical	Medium
Data rights management (ownership, storage, sharing, selling, etc.)	Technical	Medium
Risk Management Framework and Methodology	Societal	Medium

Source: CREATE-IoT.

view of their criticality that comes from an early evaluation by the European Large Scale Pilots (LSPs). This evaluation is one possible view, and it may differ if the opinion of other actors (e.g., users, service providers) is requested.

The characterization of gaps, in particular their type, their scope, the difficulties they create, and other appropriate descriptions is a first step. No listing of gaps is final and their identification will remain a work-in-progress in the IoT Standardisation community.

6.4.2 Bridging the Standardisation Gaps

As long as the gaps are existing, their resolution will have to be, one way or the other, taken into account of the IoT standardisation, in particular the SDOs/SSOs. The mapping of identified gaps on an architectural framework (such as the AIOTI HLA) creates a reference that can be understood by the IoT community and, in particular, that can be related to other frameworks e.g., those developed in other organizations, for instance in Standards Setting Organisations.

The Table 6.2 below shows a potential mapping of the above listed gaps on the AIOTI layered High Level Architecture (HLA). This is an indication

Table 6.2 Standards gaps mapped on the AIOTI HLA

Gap	Impact
Competing communications and networking technologies	Network layer
Easy standard translation mechanisms for data interoperability	IoT and application layers
Standards to interpret the sensor data in an identical manner across heterogeneous platforms	IoT layer
APIs to support application portability among devices/terminals	IoT layer
Fragmentation due to competitive platforms	Not specific to HLA
Tools to enable ease of installation, configuration, maintenance, operation of devices, technologies, and platforms	Mostly IoT layer, also Appl. and Network
Easy accessibility and usage to a large non-technical public	Not specific to HLA
Standardized methods to distribute software components to devices across a network	IoT and network layers
Unified model/tools for deployment and management of large scale distributed networks of devices	All layers; critical in IoT layer

(Continued)

Table 6.2   Continued

Gap	Impact
Global reference for unique and secured naming mechanisms	All layers
Multiplicity of IoT HLAs, platforms and discovery mechanisms	Addressed by HLA
Certification mechanisms defining "classes of devices"	Network layer
Data rights management (ownership, storage, sharing, selling, etc.)	All layers
Risk Management Framework and Methodology	All layers; interface definition

of the type of effort needed for the resolution of a gap and also of the kind of SDO/SSO that can address it in a relevant manner.

The work program of IoT standardisation is, by nature, not predictable. However, some considerations may be taken into account in order to ensure that new standards developments will foster collaboration and reduce fragmentation:

- Solutions should be transversal, with "horizontal layer" standards rather than "vertical domain" specific;
- Interoperability will be essential for the deployment of the IoT systems, to ensure seamless communication and seamless flow of data across sectors and value chains;
- New interoperability solutions should seek for integration into "horizontal" frameworks (e.g., oneM2M) rather than provide point solutions;
- Effective security and privacy solutions are key to user acceptance and should be based on global holistic approaches (e.g., security by design, privacy by design) that involve all the actors (and not just the specialists);
- Solutions are often not just technical solutions and existing standards may have to address non-technical issues.

6.5 Conclusions

As a technology, IoT is not isolated and it should work in conjunction with the development and deployment of other new technologies such as 5G, Big-Data, Cybersecurity or Artificial Intelligence. Moreover, the IoT systems should be more and more integrated with the complex systems of practically all of the very large vertical domains of today: Industry, Manufacturing, Robotics, Aeronautics, Intelligent Transport Systems, Maritime, Smart Living, eHealth, Farm & Food, Energy, Buildings, Environment, Cities, just to name a few.

The road to the Internet of Things (and even more to the Internet of Everything) is going to take time to travel. The role of IoT standardisation in the emergence of IoT on a largescale will be key. One of its major challenges is to help break the silos and support the integration of new, currently unforeseen, cross domain, federated applications based on open, interoperable solutions.

One clear lesson can be drawn already from the analysis of standards gaps and the definition of the program to address them: no single SDO or Alliance can address it with a one-fits-all solution. The large-scale deployment of a trusted and reliable IoT is going to be a global collaborative effort across standardisation organisations.

References

[1] European Commission communication on "ICT Standardisation Priorities for the Digital Single Market", COM(2016) 176 final, Brussels, 19.4.2016

[2] AIOTI WG03 Report: "IoT LSP Standard Framework Concepts Release 2.7" February 2017; https://docbox.etsi.org/SmartM2M/Open/AIOTI/

[3] AIOTI WG03 Report: "High Level Architecture (HLA) Release 2.1" September 2016; https://docbox.etsi.org/SmartM2M/Open/AIOTI/

[4] STF 505 TR 103 375 "SmartM2M IoT Standards landscape and future evolution", 10/2016. https://docbox.etsi.org/SmartM2M/Open/AIOTI/ STF505

[5] O. Vermesan and P. Friess (Eds.). Digitising the Industry Internet of Things Connecting the Physical, Digital and Virtual Worlds, Section 3.3.7., ISBN: 978-87-93379-81-7, River Publishers, Gistrup, 2016.

[6] O. Vermesan and P. Friess (Eds.). Building the Hyperconnected Society – IoT Research and Innovation Value Chains, Ecosystems and Markets, ISBN: 978-87-93237-99-5, River Publishers, Gistrup, 2015.

[7] McKinsey Global Institute, The Internet of Things: Mapping the value beyond the hype (2015), https://www.mckinsey.de/files/unlocking_the_ potential_of_the_internet_of_things_full_report.pdf

[8] AIOTI WG03 – IoT Standardisation: "Semantic Interoperability" Release 2.0 (2015), https://docbox.etsi.org/smartM2M/Open/AIOTI/!! 20151014Deliverables/AIOTI_WG3_SemanticInterop_Release_2_0a.pdf

[9] European Commission, Directorate-General for Internal Market, Industry, Entrepreneurship and SMEs: GROW – Rolling Plan for ICT Standardisation 2016, http://ec.europa.eu/DocsRoom/documents/14681/atta chments/1/translations/en/renditions/native

[10] ETSI SAREF TS 103 264 SmartM2M; Smart Appliances; Reference Ontology and oneM2M Mapping version 2.2.1 (2017), http://www.etsi.org/deliver/etsi_ts/103200_103299/103264/02.01.01_60/ts_103264v020101p.pdf

[11] ETSI TR 103 411 SmartM2M; Smart Appliances; SAREF extension investigation (2017), http://www.etsi.org/deliver/etsi_tr/103400_103499/103411/01.01.01_60/tr_103411v010101p.pdf

[12] ETSI SAREF for Energy (SAREF4ENER) TS 103 410-1 SmartM2M; Smart Appliances Extension to SAREF; Part 1: Energy Domain (2017), http://www.etsi.org/deliver/etsi_ts/103400_103499/10341001/01.01.01_60/ts_10341001v010101p.pdf

[13] ETSI SAREF for Environment (SAREF4ENVI): TS 103 410-2 SmartM2M; Smart Appliances Extension to SAREF; Part 2: Environment Domain (2017), http://www.etsi.org/deliver/etsi_ts/103400_103499/10341002/01.01.01_60/ts_10341002v010101p.pdf

[14] ETSI SAREF for Building (SAREF4BLDG): TS 103 410-3 SmartM2M; Smart Appliances Extension to SAREF; Part 3: Building Domain (2017), http://www.etsi.org/deliver/etsi_ts/103400_103499/10341002/01.01.01_60/ts_10341002v010101p.pdf

[15] Web of Things (WoT) Interest Group (IG), WoT Current Practices (2017), http://w3c.github.io/wot/current-practices/wot-practices.html

[16] STF 505 TR 103 376 "SmartM2M; IoT LSP use cases and standards gaps", 10/2016. https://docbox.etsi.org/SmartM2M/Open/AIOTI/STF505

[17] AIOTI Workshop on Security and Privacy, 16 June 2016; Final Report Workshop on Security and Privacy in IoT: https://aioti-space.org/wp-content/uploads/2017/03/AIOTI-Workshop-on-Security-and-Privacy-in-the-Hyper-connected-World-Report-20160616_vFinal.pdf

[18] Final Report European Commission of 13 January Workshop on Internet of Things Privacy and Security: https://ec.europa.eu/digital-single-market/en/news/internet-things-privacy-security-workshops-report

7

Large Scale IoT Security Testing, Benchmarking and Certification

Abbas Ahmad[1,5], Gianmarco Baldini[2], Philippe Cousin[1], Sara N. Matheu[3], Antonio Skarmeta[3], Elizabeta Fourneret[4] and Bruno Legeard[4,5]

[1]Easy Global Market, France
[2]JRC, Italy
[3]University of Murcia, Spain
[4]Smartesting Solutions & Services, France
[5]FEMTO ST/Université de Bourgogne Franche-Comté, France

Abstract

The Internet of Things (IoT) is defined by its connectivity between people, objects and complex systems. This is as vast as it sounds spanning all industries, enterprises, and consumers. The massive scale of recent Distributed Denial of Service (DDoS) attacks (October 2016) on DYN's servers that brought down many popular online services in the US, gives us just a glimpse of what is possible when attackers are able to leverage up to 100,000 unsecured IoT devices as malicious endpoints. Thus, ensuring security is a key challenge. In order to thoroughly test the internet of things, traditional testing methods, where the System Under Test (SUT) tested pre-production, is not an option. Due to their heterogeneous communication protocol, complex architecture and insecure usage context, IoTs must be tested in their real use case environment: service based and large-scale deployments.

This article describes the challenges for IoT security testing and presents a Model Based Testing approach solution, which can be used to support and EU security certification framework at European level for IoT products.

7.1 Introduction

The Internet-of-Things (IoT) is rapidly heading for large scale meaning that all mechanisms and features for the future IoT need to be especially designed and duly tested/certified for large-scale conditions. Also, Security, Privacy and Trust are critical conditions for the massive deployment of IoT systems and related technologies. Suitable duly tested solutions are then needed to cope with security, privacy and safety in the large scale IoT. Interestingly, world-class European research on IoT Security & Trust exists in European companies (especially SME) and academia where even there are available technologies that were proven to work adequately in the lab and/or small-scale pilots. More, unique experimental IoT facilities exist in the EU FIRE initiative that make possible large-scale experimentally-driven research but that are not well equipped to support IoT Security & Trust experiments.

But notably, Europe is a leader in IoT Security & Trust testing solutions (e.g. RASEN toolbox, ETSI Security TC, etc.) that can be extended to large-scale testing environments and be integrated in FIRE IoT testbeds for supporting experimentations. The ARMOUR project aims at providing duly tested, benchmarked and certified Security & Trust technological solutions for large-scale IoT using upgraded FIRE large-scale IoT/Cloud testbeds properly equipped for Security & Trust experimentations. To reach this goal, ARMOUR will:

- Enhance two outstanding FIRE testbeds ($>$ 2700nodes; $\sim$500users) with the ARMOUR experimentation toolbox for enabling large-scale IoT Security & Trust experiments.
- Deliver properly experimented, suitably validated and duly bench-marked methods and technologies for enabling Security & Trust in the large-scale IoT.
- Define a framework to support the design of Secure & Trusted IoT applications as well as establishing a certification scheme for setting confidence on Security & Trust IoT solutions.

This chapter will present first ARMOUR IoT security testing and its Model Based Testing (MBT) approach. Then, the Section 7.3 of the chapter presents the ARMOUR methodology for benchmarking security and privacy in IoT. Finally, these outcomes will be highlighted trough the prism of the on-going European wide IoT security certification process, before concluding in Acknowledgements.

7.2 ARMOUR IoT Security Testing

This section defines a common language and ontology to express and share project's objectives and activities. This includes primarily a list of vulnerabilities of IoT systems that will be addressed within the ARMOUR project, selected from the analysis of on-going IoT related security initiatives (NIST [36], oneM2M [37], OWASP IoT [34], GSMA [35]). Based on the analysis of the proposed security experiments, ARMOUR defines its methodology based on four segments of an IoT deployment:

- Devices and data
- Connectivity (Wireless)
- Platforms
- Applications and Services

The ARMOUR security framework defines the security in terms of availability, integrity and confidentiality/privacy, offering solutions and guidelines for each of the four segments and the respective identified elements to be secured (Figure 7.1).

The ARMOUR security framework takes as entry the oneM2M vulnerabilities, threats and risk assessment methodology and enriched it with missing vulnerabilities and threats based on the seven experiments to be conducted within the project.

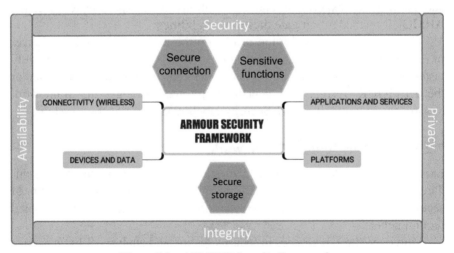

Figure 7.1 ARMOUR Security Framework.

The seven experiments (Figure 7.2) planned within the project and covering the four listed segments, are:

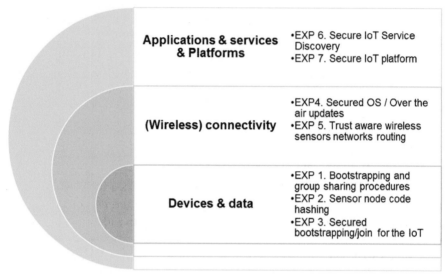

Figure 7.2   Positioning of ARMOUR experiments over IoT value chain.

For each experiment, a process is applied to identify the vulnerabilities to be addressed by the experiment and the solutions to be experimented over the ARMOUR testbed to mitigate these vulnerabilities.

This project is driven by the results on the experiments and we deliver in this chapter key elements (vulnerability and test patterns) to provide duly tested and secure solutions in the IoT domain, as well as tools, in particular the ARMOUR Testing and benchmarking methodology and framework towards the creation of an IoT Security labeling and certification system.

7.2.1 ARMOUR Testing Framework

Exisitng Machine to Machine (M2M) standards as well as emerging standards, such as oneM2M, put extreme importance into the definition of security requirements related to security functions in addition to functional requirements.

Moreover, the experiences in security testing and analysis of IoT systems, showed that their Security and Trust will depend on the resistance of the system with respect to:

- Misuses of the security functions,
- Security threats and vulnerabilities,
- Intensive interactions with other users/systems.

Based on the work performed within the project and the testing needs of each experiment, a set of security requirements and vulnerabilities that must be fulfilled by the developed systems have been identified. In order to validate them with respect to the set of requirements, three test strategies have been defined, implementable in combination or individually:

- **Security Functional testing (compliance with agreed standards/ specification):** Aims to verify that system behavior complies with the targeted specification which enables to detect possible security misuses and that the security functions are implemented correctly.
- **Vulnerability testing (pattern driven):** Aims to identify and discover potential vulnerabilities based on risk and threat analysis. Security test patterns are used as a starting point, which enable to derive accurate test cases focused on the security threats formalized by the targeted test pattern.
- **Security robustness testing (behavioral fuzzing):** Compute invalid message sequences by generating (weighted) random test steps. It enables to tackle the unexpected behavior regarding the security of large and heterogeneous IoT systems.

Model-Based Testing (MBT) approaches have shown their benefits and usefulness for systematic compliance testing of systems that undergo specific standards and that define the functional and security requirements of the system.

ARMOUR proposes a tailored MBT automated approach based on standards and specifications that combines the above mentioned three test strategies built upon the existing CertifyIt technology [16] and TTCN-3 [19] for test execution on the system under test (SUT) into one toolbox called ARMOUR Model Based Security Testing Framework. On the one hand, the CertifyIt technology has already proven its usefulness for standard compliance testing of critical systems, for instance on GlobalPlatform smartcards. Thus, building the ARMOUR approaches upon CertifyIt will allow to get the benefits of a proven technology for conformance testing and introducing and improving it in the domain of IoT. On the other hand, Testing and Test Control Notation version 3 (TTCN-3) is a standardized test scripting language widely known in the telecommunication sector. It is used by the third Generation Partnership Project (3GPP) for interoperability and certification testing, including the prestigious test suite for Long Term

Evolution (LTE)/4G terminals. Also, the European Telecommunication Standards Institute (ETSI), the language's maintainer, is using it in all of its projects and standards' initiatives, like oneM2M. Finally, this testing framework will be deployed within the ARMOUR test beds (FIT IoT lab and FIESTA), for large-scale testing and data analysis.

Based on the evaluation of the testing needs, three possible levels of automation have been identified: The ARMOUR MBT approach with automated test conception and execution based on TPLan tests and TTCN-3 scripts, manual TPLan and TTCN-3 conception and their automated execution on the SUT and finally in-house approaches for testing.

We summarize the approach and illustrate the ARMOUR MBT Security Testing Framework in Figure 7.3. It depicts the three kinds of approaches considered in ARMOUR based on the experiments needs as discussed previously: Tailored MBT approach, manual conception and in-house approach.

The MBT approach in ARMOUR Model Based Security Testing Framework relies on MBT models, which represent the structural and the behavioral part of the system. The structure of the system is modeled by UML class diagrams, while the systems behavior is expressed in Object Constraint Language (OCL) pre- and post-conditions. Functional tests are obtained by applying a structural coverage of the OCL code describing the operations of the SUT (functional requirements). This approach in the

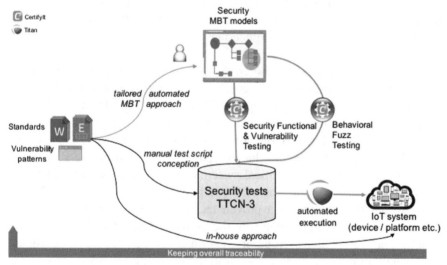

Figure 7.3   ARMOUR Model Based Security Testing Framework.

context of security testing is complemented by dynamic test selection criteria called Test Purposes that make it possible to generate additional tests that would not be produced by a structural test selection criterion, for instance misuse of the system (Model-Based Security Functional Testing) and vulnerability tests, trying to bypass existing security mechanisms (Model-Based Vulnerability Testing). These two approaches generate a set of test cases stored inside a database and then executed on the system. To the difference of them, robustness testing in our context, based on the same model, will generate randomly a test step based on the same MBT model by exercising different and unusual corner cases on the system in a highly intensive way, thus potentially activating an unexpected behavior in the system.

7.2.2 Identified Security Vulnerabilities and Test Patterns

ARMOUR project proposes a security framework based on risk and threat analysis that defines a list of potential vulnerabilities for the different IoT layers (Table 7.1). Each vulnerability is associated to a test pattern. To define the test patterns for each vulnerability, a set of test procedures have been conceived for verifying its system resistance to the vulnerabilities of its experiments. Based on these test procedures specific for the ARMOUR experiments that are introduced in the following section, we have created a library of test patterns generalized to the four IoT segments. The following Figure 7.4. illustrates the methodology applied for the definition

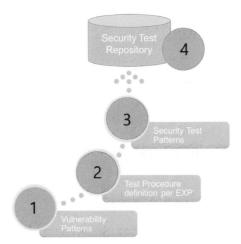

Figure 7.4   Security Test Patterns definition methodology.

Table 7.1 List of defined vulnerabilities

Id	Title
V1	Discovery of Long-Term Service-Layer Keys Stored in M2M Devices or M2M Gateways
V2	Deletion of Long-Term Service-Layer Keys stored in M2M Devices or M2M Gateways
V3	Replacement of Long-Term Service-Layer Keys stored in M2M Devices or M2M Gateways
V4	Discovery of Long-Term Service-Layer Keys stored in M2M Infrastructure
V5	Deletion of Long-Term Service-Layer Keys stored in M2M Infrastructure equipment
V6	Discovery of sensitive Data in M2M Devices or M2M Gateways
V7	General Eavesdropping on M2M Service-Layer Messaging between Entities
V8	Alteration of M2M Service-Layer Messaging between Entities
V9	Replay of M2M Service-Layer Messaging between Entities
V10	Unauthorized or corrupted Applications or Software in M2M Devices/Gateways
V11	M2M System Interdependencies Threats and cascading Impacts
V12	M2M Security Context Awareness
V13	Eaves Dropping/Man in the Middle Attack
V14	Transfer of keys via independent security element
V15	Buffer Overflow
V16	Injection
V17	Session Management and Broken Authentication
V18	Security Misconfiguration
V19	Insecure Cryptographic Storage
V20	Invalid Input Data
V21	Cross Scripting

of generalized security test patterns (3) and towards the conception of security test cases (4).

ARMOUR security test patterns define test procedures for verifying security threats of IoT systems, representatives of different IoT levels, thus facilitating the reuse of known test solutions to typical threats in such systems.

Table 7.2 presents the test patterns and gives an overview of the vulnerabilities that they cover.

With the test vulnerabilities and test patterns defined, each of the testing use cases should generate a set of test cases in order to assess their results. The global overview of the testing methodology is presented in the next section.

7.2.3 ARMOUR IoT Security Testing Approach

The overall ARMOUR test environment for IoT security testing is the following:

Table 7.2 Vulnerabilities overview

Test Pattern ID	Test Pattern Name	Related Vulnerabilities
TP_ID1	Resistance to an unauthorized access, modification or deletion of keys	V1, V2, V3, V4, V5
TP_ID2	Resistance to the discovery of sensitive data	V6
TP_ID3	Resistance to software messaging eavesdropping	V7
TP_ID4	Resistance to alteration of requests	V8
TP_ID5	Resistance to replay of requests	V9
TP_ID6	Run unauthorized software	V10
TP_ID7	Identifying security needs depending on the M2M operational context awareness	V12
TP_ID8	Resistance to eaves dropping and man in the middle	V13
TP_ID9	Resistance to transfer of keys via of the security element	V14
TP_ID10	Resistance to Injection Attacks	V16
TP_ID11	Detection of flaws in the authentication and in the session management	V17
TP_ID12	Detection of architectural security flaws	V18
TP_ID13	Detection of insecure encryption and storage of information	V19
TP_ID14	Resistance to invalid input data	V20

The first step after vulnerability and test pattern identification, is to build a Model Based Testing model representing the behavior of the System Under Test (SUT). The model takes as input a set of security test patterns in order to generate security tests in TTCN-3 format. Next, the tests are compiled and it produces an Abstract Test Suite (ATS) in TTCN-3 format. In order to have an executable ATS, we need to use a compiler that will transform the TTCN-3 code in an intermediate language, like C++ or Java. This is done on purpose because the TTCN-3 code is abstract in a sense that it doesn't define the way to communicate with the SUT or how the TTCN-3 structures will be transformed to a real format. This is a task for the System Adapter and the Codec. There are few commercial and non-commercial compilers available for download from the official TTCN-3 webpage. Usually, all of the compilers follow this procedure:

- Compile TTCN-3 to a target language (Titan uses C++, TTWorkbench uses Java),
- Add the Codec and System Adapter (SA) written in the target language.

In the scope of ARMOUR, Titan test case execution tool was used. The goal of the testing process, as described in the preceding Figure 7.5. ARMOUR overall test environment is to execute the MBT-generated tests on the SUT.

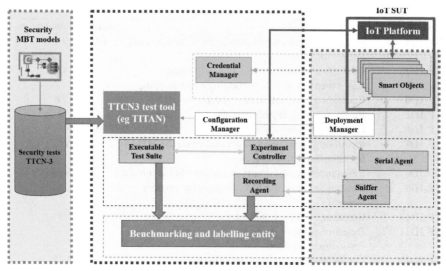

Figure 7.5   ARMOUR overall test environment.

System under test (SUT) refers to a system that is being tested for correct operation. According to ISTQB it is the test object. The term is used mostly in software testing. A special case of a software system is an application which, when tested, is called an application under test. The SUT can be anything involved in an IoT deployment for example:

- Secure communication
- IoT platform
- Device software
- Device Hardware

In ARMOUR, two types of SUT are taken into consideration, IoT Platform and IoT smart objects (Devices). All tests produced are here to stimulate the SUT's and get results in order to assert them and setup a label.

We use for EXP7 an OM2M IoT platform as SUT for the test system. OM2M is an oneM2M standard implementation. The choice of OM2M was made with regards that it is open source and accessible via Eclipse repositories. For a general usage of ARMOUR methodology, any IoT platform implementation is compatible.

SUT can also be called TOE, a Target of Evaluation, a term coming from the Common Criteria. It is defined as a set of software, firmware and/or hardware possibly accompanied by guidance. While there are cases where a

TOE consists of an IT product, this need not be the case. The TOE may be an IT product, a part of an IT product, a set of IT products, a unique technology that may never be made into a product, or a combination of these.

As for the devices tested in ARMOUR, they are devices made available by the FIT IoT- LAB. The test environment is composed of 2 main parts:

I) A laboratory test environment

The first part of the test environment is the laboratory environment. It is composed of many software applications that must be made available online on a cloud machine or other physical address.

- **Credential Manager**: The authority that distributes Cipher keys or certificates.
- **TTCN-3 test tool** (eg TITAN), where the automated tests are stored, compiled and run on the SUT via the Experience controller (Shown in Figure 7.5).
- **Experience Controller** (EC) is a bridge between the Titan tests and FIT IoT lab test environment. It is adaptable to accept other testbeds.
- **Recording Agent** or **sniffer** must be locally installed on the testbed in order to listen to nodes messages. The recorded messages are made accessible online on the cloud.

II) A large-scale test bed: The IoT-LAB

IoT-LAB provides a large-scale test bed. It provides full control of network nodes and direct access to the gateways to which nodes are connected, allowing researchers to monitor nodes energy consumption and network-related metrics, e.g. end-to-end delay, throughput or overhead. The facility offers quick experiments deployment, along with easy evaluation, results collection and analysis. Defining complementary testbeds with different node types, topologies and environments allows for coverage of a wide range of real-life use cases.

IoT-LAB testbeds are located in six different sites across France which gives forward access up to 2700 wireless sensors nodes: Inria Grenoble (928), Inria Lille (640), ICube Strasbourg (400), Inria Saclay (307), Inria Rennes (256) and Institut Mines-Télécom Paris (160) (Numbers as of May 2017).

The IoT-LAB hardware infrastructure consists of a set of IoT-LAB nodes. A global networking backbone provides power and connectivity to all IoT-LAB nodes and guaranties the out of band signal network needed for command purposes and monitoring feedback. This test bed facilitates the availability and implementation of devices in order to run the security tests in a large-scale environment.

7.2.4 Large Scale End to End Testing

Internet of Things (IoT) applications can be found in almost all domains, with use cases spanning across areas such as healthcare, smart homes/ buildings/cities, energy, agriculture, transportation, etc.

It is impossible to provide an exhaustive list of all application domains of the IoT. IoT systems involved in a solution has in most common cases a wide spread of different components such as Hardware (Devices, Sensors, Actuators...), Software and Network protocol. ARMOUR project introduces different experiences to test the different components. More specifically, each experience proposes individual security tests covering different vulnerability patterns for one or more involved components. In ARMOUR, EXP1 covers the security algorithms for encryption/decryption and protocols for secure communication. On the other hand, EXP7 covers the IoT data storage and retrieval in an IoT platform. Both experiences can wisely be integrated together. Indeed, it makes sense to pair the secure transfer and storage of data with an IoT platform thus making sure that the confidentiality and integrity of the data is maintained from the moment it is produced by a device, until it is read by a data consumer.

The Model for End-to-End security is conceived as follows:

A wise integration of the experiments is made from different experience models. We call it an Integration model which is used in combination of security vulnerability scenarios in order to generate Abstract Test Cases (Figure 7.6).

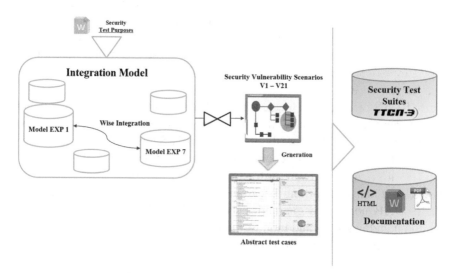

Figure 7.6   End-to-End security description.

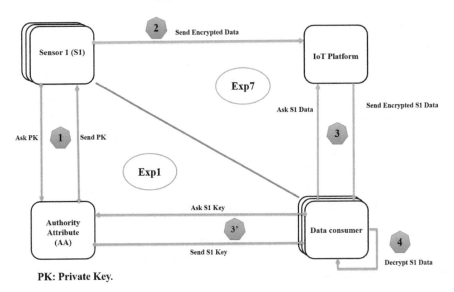

Figure 7.7   Large-Scale End-to-End scenario.

The test cases then follow the normal MBT procedure that is test publication as executables for test automation or as documentation for manual testing.

The End-to-End scenario (E2ES) is described in the above Figure 7.7.

Both data consumer/producers retrieve keys from a credential manager and the IoT platform role is to support the data storage/retrieval. However, a single security problem in any of the components involved in this scenario can compromise the overall security. There is a perception that end-to-end security is sufficient as a security solution, and that network-based security is obsolete in the presence of end-to-end security. In practice, end-to-end security alone is not sufficient and network-based security is also required. This demonstrates that the End-to-End security scenario is here to complete the experience's tests and is not an exhaustive testing method.

7.3 ARMOUR Methodology for Benchmarking Security and Privacy in IoT

Moving from isolated IoT domains to real large-scale deployments requires a better understanding and consensus about actual implications of envisioned infrastructures. For example [1]:

- Is the deployment doing what it is supposed to do?
- Is it the best way to achieve the expected goal?
- Is the deployment creating unexpected issues?
- Is the deployment achieving the expected economic performance?

Providing answers to these questions is of interest to people involved in IoT deployments to identify good practices, avoid traps and make good choices. In this sense, the benchmarking of security and trust aspects is crucial to guarantee the success of large-scale IoT deployments.

Benchmarking IoT deployments should meet different objectives. On the one hand, as we are still in a disharmonized landscape of technologies and protocols, for experimenters and testers, having the whole vision of difficulties and opportunities of IoT deployments is quite difficult. Consequently, there is a need to follow the deployment process, to understand how deployment works, and which security and trust technologies are involved, as well as the impact of a security flaws related to such technologies. On the other hand, and also related with such fragmented landscape of security solutions for the IoT, large-scale deployments will be based on a huge amount of different technologies and protocols. Therefore, the identification of common metrics is crucial to be able to assess and compare such solutions in order to identify good security and trust practices and guidelines.

Towards this end, one of the main goals of the project stems from the need to identify a suitable benchmarking methodology for security and trust in IoT, as a baseline for the certification process. This methodology, as shown in next sections is based on the identification of different metrics associated to different functional blocks, in order to benchmark security and privacy on the different experiments designed in ARMOUR. Specifically, micro-benchmarks provide useful information to understand the performance of subsystems associated with a smart object. Furthermore, they can be used to identify possible performance bottlenecks at architecture level and allow embedded hardware and software engineers to compare and asses the various design trade-offs associated with component level design. Macro-benchmarks provide statistics relating to a specific function at application level and may consist of multiple component level elements working together to perform an application layer task.

The ARMOUR project is focused on carrying out security testing and certification in large-scale IoT deployments. Between both processes, benchmarking is intended to provide different results from testing to serve as the baseline for the certification scheme. Towards this end, for both micro-benchmarking and macro-benchmarking, this methodology will be based on the identification of different metrics per functional block (e.g. authentication). Benchmarking results will help to increase the trust level on the assurance of security properties in IoT products and solutions. Based on this, the next section provides a description of the ARMOUR benchmarking methodology.

7.3.1 Approach Overview

A high-level overview of the proposed benchmarking methodology is shown in Figure 7.8, depicting, the methodology comprising five main stages:

- Experiment design. A concrete experiment is defined in the context of a specific IoT application scenario or use case.
- Test design. The description of the experiment is used to identify different threats and vulnerabilities that can be tested over it. These vulnerabilities and threats are identified from a set of already established vulnerability patterns, and used as an input to generate test patterns in order to define the procedure to test a specific threat.
- Test generation. Based on test patterns, a test model formalizes a specific subset of the functionality of the experiment. Using these models, a set of test suites are generated to be then executed.
- Test execution. From the set of test suites, a set of test adapters are defined to guarantee suites can be executed under different environments and conditions. Specifically, in the context of the project, these tests are intended to be executed in both in-house scenarios and large-scale setting, through the use of FIT IoT-Lab[1].

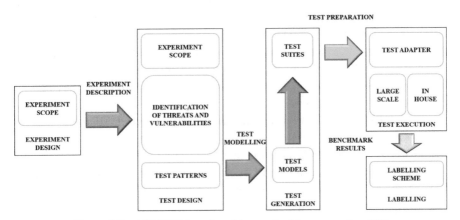

Figure 7.8 ARMOUR benchmarking methodology overview [23].

---

[1] www.iot-lab.info

- Labelling. The results from the execution of these tests are the used during the certification process by making use of a security labelling scheme.

In following sections, a more detailed overview of these stages is provided.

7.3.2 Experiment Design

The ARMOUR project considers a large-scale experimentally-driven research approach due to IoT complexity (high dimensionality, multi-level interdependencies and interactions, non-linear highly-dynamic behavior, etc.). This research approach makes possible to experiment and validate research technological solutions in large-scale conditions and very close to real-life environments. As a baseline for the benchmarking methodology, different experiments have been designed to test different security technologies, protocols and approaches that are intended to be deployed in typical IoT scenarios (introduced as well in Figure 7.2):

- EXP 1: Bootstrapping and group sharing procedures
- EXP 2: Sensor node code hashing
- EXP 3: Secured bootstrapping/join for the IoT
- EXP 4: Secured OS/Over the air updates
- EXP 5: Trust aware wireless sensors networks routing
- EXP 6: Secure IoT Service Discovery
- EXP 7: Secure IoT platforms

This set of experiments constitutes the basis to obtain benchmarking results from the methodology application to such scenarios. In particular, these experiments have been designed to test specific security properties of technologies and protocols, such as the Datagram Transport Layer Security (DTLS) [2], as the main security binding for the Constrained Application Protocol (CoAP) [3]. These experiments also involve the use of different cryptographic schemes and libraries that will be tested, in order they can be certified according to benchmarking results. By considering different security aspects, during this stage, following a similar approach to the EALs in Common Criteria [4], a risk analysis is performed regarding the assurance level for the environment where the device will operate.

The design and description of such experiments is based on the definition of different required components and testing conditions, to ease the test design, modelling and execution that are defined in the following subsections.

7.3.3 Test Design

During test design, experiments are set up and prepared through the specification of security test patterns that describe the testing procedures. By considering the set of different experiments, as well as the security protocols and technologies involved in each of them, this stage is focused on the definition of different tests to assess the fulfilment of specific security properties on those experiments. These properties are:

- **Authentication**. The devices should be legitimate.
- **Resistance to replay attacks**. Intermediate node can store a data packet and replay it at a later stage. Thus, mechanisms are needed to detect duplicate or replayed messages.
- **Resistance to dictionary attacks**. Intermediate node can store some data packets and decipher them by performing a dictionary attack if the key used to cipher them is a dictionary word.
- **Resistance to DoS attacks**. Several nodes can access to the server at the same time in order to collapse it. For this reason, the server must have DoS protection or a fast recovery after this attack.
- **Integrity**. Received data are not tampered with during transmission; if this does not happen, then any change can be detected.
- **Confidentiality**. Transmitted data can be read only by the communication endpoints.
- **Resistance to MITM (man-in-the-middle) attacks**. The attacker secretly relays and possibly alters the communication. The endpoints should have mechanism to detect and avoid MITM attacks.
- **Authorization**. Services should be accessible to users who have the right to access them.
- **Availability**. The communication endpoints should always be reached and should not be made inaccessible.
- **Fault tolerance**. Overall service can be delivered even when a number of atomic services are faulty.
- **Anonymization**. Transmitted data related to the identity of the endpoints should not be sent in clear.

Such properties have been extracted from some of the most referenced security aspects that can be found in current IoT literature [5–9]. These properties can be security properties, such as authentication and integrity, or resistances to certain attacks, such as the MITM and the replay attack.

The proposed methodology is based on the use of these properties and the set of identified vulnerabilities in ARMOUR D1.1 [10] (Table 7.1) that are extracted from oneM2M vulnerabilities [11]. From these properties and the set of vulnerabilities that are described, different tests can be designed in order to prove previous experiments are actually satisfying the set (or a subset) of security properties.

It should be pointed out that the identified security properties are intended to serve as a starting point to build a more generic, stable and holistic approach for security certification in IoT scenarios. This set will evolve during the project to cover other security vulnerabilities.

From these security vulnerabilities and properties, security test patterns are identified to define the testing procedure for each of them. In the context of the project, D2.1 [12] (Table 7.2) already identifies a set of test patterns to be considered for the seven experiments. With all this information, tests patterns can be designed by associating security vulnerabilities and properties. These tests are defined following the schema of the Figure 7.9, which includes a description for each field. The main part of the definition of the test is the test description that must include the steps that we have to follow to make the test and when the test is satisfactory or not.

The proposed template is intended to identify the association or relationship between the test patterns with the security properties and vulnerabilities. The identification of such relationship aims to foster the understanding of the benchmarking methodology approach towards a security certification scheme for the IoT.

Test pattern ID	The identifier of the test pattern
Stage	It refers to the specific stage or step of an experiment
Protocol	The technology or protocol related to the test pattern
Property tested	The security property related to the test pattern
Test diagram	A figure with the main components involved in the test pattern
Test description	Description of the steps and conditions related to the test pattern
References	Vulnerabilities related to this test pattern

Figure 7.9 Test pattern template associating security vulnerabilities and properties [23].

7.3.4 Test Generation

From the test patterns that are designed in the previous stage, during test generation, real tests are defined in order to validate different security properties. For testing purposes, different strategies can be employed; indeed, software testing is typically the process for verifying different requirements are fulfilled regarding the expected behavior of a System Under Test (SUT). In this sense, software security testing [13] is intended to validate that security requirements are satisfied on a specific SUT. According to [14], security testing techniques can be deployed according to their test basis within the secure software development lifecycle into four different types:

- Model-based security testing that is related to the requirements and design models that are created during the analysis and design.
- Code-based testing, which is focused on source and byte code created during development.
- Penetration testing and dynamic analysis on running systems, either in a test or production environment.
- Security regression testing performed during maintenance.

For the proposed methodology, ARMOUR is based on the use of the Model-Based Testing (MBT) approach [15] (as the generalization of Model-Based Security Testing (MBST)). MBT is mainly based on the automatic generation of test cases from a SUT, so consequently, MBST aims to validate security properties with a model of the SUT, in order to identify if such properties are fulfilled in the model.

7.3.5 Test Execution

Once suitable tests are generated, they must be adapted [20] in order to be executed on different environments with its own interfaces and behavior. In the context of the project, generated tests can be executed on an in-house or external large-scale testbed. Both testbed approaches aim to serve as platforms to generate benchmark results from the execution of security tests. In particular, during test execution, following tasks can be carried out [21]:

- Measure, in order to collect data from experiments.
- Pre-process, for obtaining "clean" data to ease assessment and comparisons among different security technologies and approaches.
- Analyse to get conclusions from benchmarking results as a previous step for the labelling and certification.
- Report, informing the inferred conclusions from experiments results.

As an in-house environment example, PEANA is an experimentation platform based on OMF/SFA. An experimenter intended to use this platform will, in first place, use the web portal to schedule and book its experiment. After booking the required components for his experiment, the platform will allow him to access the Experiment Controller, via SSH. This is the place where the experiment will take place. This way, the experimenter must define his experiment using the OMF Experiment Description Language (OEDL) [22] syntax indicating which are the Resource Controllers to be used, i.e. the entity responsible for managing the different scheduled IoT devices. Such definition can also include the new firmware to be tested, and the gathering of returned information. After executing the experiment, the experimenter will be allowed to analyse the performance of his experiment according to the obtained results.

7.3.6 Labelling

During this last stage of the methodology, benchmarking results from the previous stage are used as an input for labelling and certification purposes. The main purpose of this stage is to check if theoretical results that were expected during initial stages are obtained after test executions.

It implies the design a labelling scheme specifically tailored to IoT security and trust aspects, so different security technologies and approaches can be compared, to certify different security aspects of IoT devices. The establishment of this scheme is key to increase trust of IoT stakeholders for large-scale deployments. However, labelling approaches for IoT need to consider the dynamism of IoT devices during their lifecycle (e.g. operating on changing environments), which makes security certification a challenging process.

As an initial approach of labelling for IoT security, below we provide a description of the main aspects that are currently being considered. These aspects will be enhanced within next deliverables to build a solid labelling approach for security aspects in IoT environments. For this approach, it should be noted that labelling must be taken into account the context of the scenario that is being tested and the certification execution. For this reason, and based on Common Criteria (CC) approach, three mains aspects will be considered to be included in the label:

- TOE (Target of Evaluation): In CC, a TOE is defined as a set of software, firmware and/or hardware possibly accompanied by guidance. In this case, a TOE is defined as a context, for example health, industry or home automation. The TOE includes all the devices of the context.

For example, in home automation, the TOE could be the fire sensor, the bulbs, the washing machine, etc.

- Profiles (level of protection): Low, medium and high. The level of protection is related to the threats that can be avoided in the tested scenario.
- Certification execution: The proposed certification execution has 4 levels, which are shown in Figure 7.10. This aspect is intended to be further extended by the certification approach.

In order to be able to label a scenario, we have to consider some metrics that are going to be associated to the execution of the tests, such as detectability, impact, likelihood (that includes difficulty, motivation and benefits), difficulty in obtaining a valid message, percentage of the communication protected with integrity, recoverability, percentage of requests per second necessary, sensibility of the data, time, key length, the facility to hack the server, etc.

In this way, based on different metrics, we have different levels of security associated to a mark for a subset of the security properties that are identified in Section 7.3.3. This association between security properties and marks is shown in Figure 7.11.

As already mentioned, the context must be taken into account in the labelling process. For this reason, we consider an additional parameter: Risk. This parameter is the product of Likelihood and Impact. Likelihood is related to the probability of happening, taking into account the benefit that a person could obtain hacking it, whereas impact is related to the damage that is produced in the scenario in terms of money, sensitive data filtered, scope, etc. These parameters can be variable for each vulnerability. The levels of each one is shown in the Figure 7.12.

Figure 7.10 Certification execution levels [23].

Authentication Client/Server	0	Mutual and strong
	1	Strong server, weak or without authN client
	2	Strong client weak or without authN server
	3	Weak/without authN
Resitance to Replay attacks	0	Protected
	1	Non protected but a valid message cannot be obtained
	2	Non protected and a valid message can be obtained with difficulty/weak protection
	3	Non protected, it can be obtained easily
Resistance to Dictionary attacks	0	Non applicable
	1	Strong key
	2	Weak key
Integrity	0	Total
	1	Partial
	2	None
Resistance to DoS attacks	0	Minimun state
	1	Big state
Confidentiality	0	Total with secure encryption
	1	Partial with secure encryption
	2	Total with insecure encryption
	3	Partial with insecure encryption
	4	None
Resistance to MITM attacks	0	Detectable
	1	Non detectable

Figure 7.11 Association between security properties and marks based on metrics [23].

Likelihood	0	Null benefit
	1	Medium benefit
	2	High benefit
Impact	0	Little damage and recuperable
	1	Limited damage (Scope, monetary losses, sensible data...)
	2	High damage

Figure 7.12 Marks for likelihood and impact parameters [23].

Once we have given each threat and scenario a score (risk and mark), we can define the TOE and the profiles, specifying the minimum level of security they must have, the minimum score (in terms of risk and mark) for each property. If the scenario we are testing achieves the level of security of several profiles, it will be labelled with the major level of security. This implies that an upper level satisfies the requirements of the lower levels.

7.4 A European Wide IoT Security Certification Process

7.4.1 Needs for a European Wide IoT Security Certification Process

On the basis of the concepts described in the previous Sections 7.2, 7.3 and this section, we describe a potential framework for IoT security certification at European level, which is able to support the testing of security and privacy requirements and address the limitations of Common Criteria.

The need for a harmonized security European certification scheme has already been suggested by various studies including [24] and [25], where it is pointed out that current national certification schemes (e.g., Germany, UK and France) could form the basis to create a European certification scheme based on a common approach.

A common European certification scheme would bring a higher level of cyber-confidence to industry buyers and users. A harmonized IoT certified device and product at European level could become an added value in the market. As described in [24], a harmonized security EU certification may be more widely recognized as an international 'quality label' and, hence, support the international competitiveness of European producers. Meanwhile, non-European producers that obtained the same European certification would benefit in an equal way from this 'quality label'.

At the same time, the limitations of existing security certification process like Common Criteria should be addressed. These limitations have been already identified in literature and they are briefly summarized here:

- *Re-certification and patching.* Re-certification of an already certified system or product is an issue raised in [26] and [27] for Common Criteria. Security certification is usually done against a static version of the product and its operative environment. As described before, the IoT environment is especially characterized by dynamic changes of the product and its configuration (in some cases due to the need to install security patches). We note that some European countries like France have already proposed alternative approaches for security certification like the Certification de Sécurité de Premier Niveau (CSPN) described in [28].
- *Lack of mutual recognition.* Security certification profiles and testing defined in some European countries may not be equivalent to security profiles in other countries. The lack of mutual recognition is disruptive to the European single digital market, because consumers of security certified products will not be able to compare the levels of security

certification. Existing European organizations like SOG-IS (Senior Officials Group for Information System Security) are already addressing this issue by working on the harmonization of protection profiles at European level.

- *Certification costs.* Common criteria certification is considered a long and expensive process, which does not make it suitable for fast market deployment or relative short product cycles as in the consumer market [29] and [30]. This issue is particularly relevant for IoT products and devices because of the low profit margins for IoT manufacturers.

The proposed European wide framework aims to address and mitigate the previous issues on the basis of the concepts and tools (MBT, TTCN) described in the previous sections of this chapter.

7.4.2 Main Elements of the European wide IoT Security Certification Process

The overall representation of the framework is provided in Figure 7.13. The green elements represent artefacts (e.g., test suites) while the azure rectangles represent the roles. The main roles and elements in Figure 7.13 are described here:

- European Governing Board. This is the main governing body of the overall EU security certification framework. The European Governing board is composed at least by the representatives of the national certification bodies and the European Commission. SOG-IS will also be part of the European Governing Board. The EGB is responsible for drafting and managing changes to the security certification process. The EGB is also responsible for defining and maintaining the benchmarks.
- IoT Product Manufacturer. This is the manufacturer of the product to be submitted for certification. Manufacturers can be present in different domains or a single domain (e.g., road transportation or energy). The IoT product manufacturer is also responsible to express needs, identify vulnerabilities or define security requirements.
- European accreditation bodies and auditors. They are responsible for the accreditation of the certification centres and the periodic auditing.
- The Labelling Program Authority is the European (or member state entity), which is responsible for assigning the labels after a successful certification process. The Labelling Program authority associates the certification environment, test suites, tools, domain and processes to

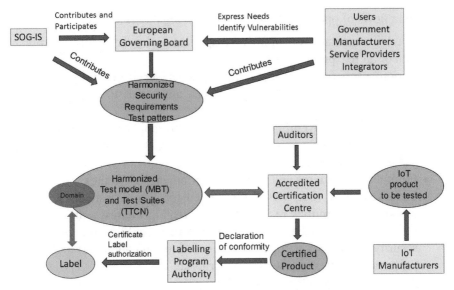

Figure 7.13 Proposed IoT EU Security Certification Framework.

specific labels, which are then provided to the IoT manufacturer after a product submission.

- Accredited certification centres. An accredited certification centre performs the test execution on the basis of the pre-defined harmonized protection profile. Note that existing test beds could be used. A number of firms in Europe have been certified as Commercial Licensed Evaluation Facilities (CLEFs) under the Common Criteria, and do certification work that is recognised across participating states. They mostly evaluate software products for government use, though there is some work on products such as smartcards that are sold into the banking industry.

- Users. They are the users of the certified product. They use the label information as a metric to drive their procurement process. Users can be citizen, public (e.g., government) or private companies. Together with manufacturers, governments, service providers and integrators, they express needs, identify vulnerabilities or define security requirements.

- SOG-IS is the European security organization described before. We consider SOG-IS a very important element of the European security certification framework because of its experience and capabilities. Still, the role and duties of SOG-IS in such a framework must be carefully evaluated together with SOG-IS representatives.

The main flows of operation in the framework are the following:

- The IoT manufacturer submits the application for the security certification of an IoT product.
- Various users (manufacturers, governments, service providers, integrators) express the security requirements and needs to define new test patterns and test suites.
- The European governing board review the applications for new test patterns and security requirements and ensure that they are harmonized at European level.
- From the test patterns and security requirements, test suites (i.e., TTCN) and test models (i.e., MBT) are created for specific categories of IoT products. The entities responsible for creating the test suites and test models is not defined at this stage, but they could be represented by security companies with skills in security certification and drafting of protection profiles from Common Criteria.
- A label is issued by the labelling program authority once that an accredited certification centre has successful certified an IoT product.

7.4.3 Security and Privacy Requirements

One task, which can be performed by the proposed EU IoT security certification framework is to address privacy requirements as well.

The concept of privacy certification is not new, even if security certification (or safety certification) has been historically the main priority. European Commission's General Data Protection Regulation [31] in Recital 77 encourages the "establishment of certification mechanisms, data protection seals and marks" to enhance transparency, legal compliance and to permit data subjects [individuals] the means to make quick assessments of the level of data protection of relevant products and services.

A relevant case study for Privacy certification is the concept of Privacy Seal [32]. The Privacy seal is a trans-European privacy trust mark issued by an independent third party certifying compliance with the European regulations on privacy and data protection. See (see https://www.european-privacy-seal.eu/ by EuroPriSe for more information on the Privacy Seal and the activities carried out by EuroPriSe. The Privacy seal concept is relatively similar to the label concept of security certification where the label is the seal itself.

The overall process to obtain a Privacy Seal could also be similar to envisaged EC security certification process described in the previous section.

Private and public manufacturers of IoT products can apply for the certificate and related European label. The trust mark is awarded after successful evaluation of the product or service by independent experts and a validation of the evaluation by an impartial certification authority.

Reference [32] provides and extensive description of the most common Privacy Certification processes available in the world. One of the main examples is TRUSTe, which defines processes for Privacy certifications for various products and services. In [33] are defined Privacy certification standards for Smart Grids, Enterprise and others. TRUSTe works closely with stakeholders to identify the needs for the definition of new Privacy certification standards. The standards define the Privacy Program requirements, the vendor must satisfy in its service or product. Examples of requirements defined in the TRUSTe standards are related to protection against phishing or the implementation of encryption methods for data protection and data confidentiality.

These examples already show that security certification and privacy certification cannot be disjointed but they should be combined as they often address the same or similar requirements (e.g., access control, confidentiality) or solutions (e.g., cryptographic algorithms).

Figure 7.14 provides a preliminary description on the potential process to support both security and privacy requirements. While applications experts (the various users from the previous subsection) could define the security requirements, the EDPS could work with the EGB and other stakeholders to define the privacy requirements. Both categories of requirements can be used to define the draft test patterns and models. Then the flow could be similar to what already presented in Figure 7.13.

7.4.4 How the EU Security Certification Framework Addresses the Needs

In this section, it is described how the proposed EU security certification framework could address the needs and limitations of common criteria identified in Section 7.4.1.

The *Lack of mutual recognition* is addressed by defining harmonized test patterns, models and suites at European level. While, these test artefacts could be still specific for domain (e.g., specific context like automotive, energy), they would be harmonized at European level for classes of products (e.g., the ITS DSRC system in a vehicle or an IoT Gateway in a smart home).

The *Re-certification and patching* issue is addressed by the use of models and well-defined test suites. By using a common criteria terminology where

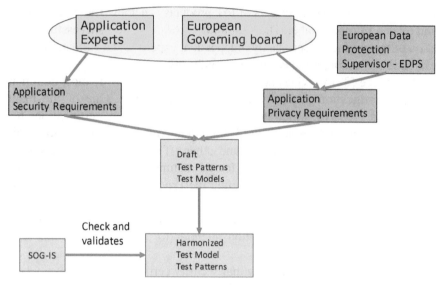

Figure 7.14   Combination of security and privacy requirements.

the TOE is the IoT device to be certified, the ARMOUR project tackles finely the issue on the re-certification and helps the review of security impact using a safe regression testing techniques. Safe regression testing techniques make possible to select test cases for execution relevant for testing the modified TOE without omitting test cases that may reveal faults. The re-evaluation of a TOE is based on a comparison between the initial and the updated version of the TOE's evidence. Several elements being part of the TOE's evidence may evolve:

- The model (the TOE's functional specification has evolved)
- The security patterns (new security properties are tested for the same TOE, for instance to reach a higher label)
- Both may evolve (as the changes in the security test patterns are often due to the development of new functionalities in the TOE)

The test tool based on an automated comparison of the TOE's formal evidence expressed in the form of an MBT model) can detect if there are any impacted security properties. Based on the analysis it will:

- Select a set of existing test cases that are impacted by the updated TOE's evidence
- Generate new tests to cover any new or evolved security test pattern

- Report on impacted vulnerability patterns, which ensure traceability to the security properties part of the security certification.

These output elements of the ARMOUR Testing Framework will serve as facilities to easy and lower the cost of the re-certification. The systematic and automated process offered by the framework will on the one hand easy the decision process on the need for re-certification, as it will report on the impacted security test cases and thus impacted security properties. On the other hand, only a sub-set of the test cases will be executed on the test bed and analysed, which lowers the processing and post processing testing activity.

The *Certification costs* issue is mitigated (not resolved as described below) by the analysis provided above (i.e., use of automated test suites) and by the consideration that a single European combination of test model, pattern and suite is created for categories of IoT devices thus creating a mass market for security certification for specific categories of IoT products.

The trade-off in setting up the EU Security certification framework is the cost of setting up this framework with an increased role for SOG-IS, the creation of the EGB and the definition of new flows and processes. On the other side, these processes and roles (e.g., accredited auditor and security certification lab) do already exist in Europe but they are quite fragmented. The authors of this book chapter believe that the adoption of a common methodology and tools can improve the security and privacy certification process in Europe.

7.5 Conclusion

This book chapter has described the application of Model Based Testing (MBT) for the security certification of IoT products. The application of MBT can improve the efficiency of the IoT security certification process, which is especially important in the IoT context, due to the presence of a wide range of vulnerabilities and the dynamic environment where IoT products are used.

The book chapter does also describe a potential EU-wide security certification framework for IoT security certification, with a clear definition of roles and processes.

Acknowledgements

This project has received funding from the European Union's Horizon 2020 research and innovation programme under grant agreement No 688237.

References

[1] PROBE-IT, Deliverable D2.1, 'Benchmarking framework for IoT deployment evaluation', Oct, 2012.

[2] E. Rescorla, N. Modadugu, 'RFC 6347 – Datagram Transport Layer Security Version 1.2.', Internet Engineering Task Force (IETF), Jan, 2012.

[3] Z. Shelby, K. Hartke, C. Bormann, 'RFC 7252 – The Constrained Application Protocol (CoAP)'. Internet Engineering T ask Force (IETF), June, 2014.

[4] The Common criteria, 'Collaborative Protection Profiles (cPP) and Supporting Documents (SD)', 2015.

[5] G. Selander, F. Palombini, K. Hartke, 'Requirements for CoAP End-To-End Security', 2016.

[6] K. Moore, R. Barnes, H. Tschofenig, 'Best Current Practices for Securing Internet of Things (IoT) Devices', 2016.

[7] M. Abomhara, G. Koien, 'Cyber Security and the Internet of Things: Vulnerabilities, Threats, Intruders and Attacks', Journal of Cyber Security Vol. 4, pp. 65–88, 2015.

[8] T. Heer, O. Garcia-Morchon, R. Hummen, S. Loong, S. Kumar, K. Wehrle, 'Security Challenges in the IP-based Internet of Things', Wireless Pers Commun, Vol. 61, pp. 527–542, 2011.

[9] H. Suo, J. Wan, C. Zou, and J. Liu, 'Security in the Internet of Things: A Review', International Conference on Computer Science and Electronics Engineering, 2012.

[10] ARMOUR, 'Deliverable. D1.1 – ARMOUR Experiments and Requirements', Aug, 2016.

[11] OneM2M-TR-0008, 'Analysis of Security Solutions for oneM2M System v0.2.1', 2013.

[12] ARMOUR, 'Deliverable D2.1 – Generic test patterns and test models for IoT Security Testing', Aug, 2016.

[13] B. Potter, G. McGraw, 'Software security testing', IEEE Security & Privacy, Vol 2, pp. 81–85, 2004.

[14] M. Felderer, M. Büchler, M. Johns, A. Brucker, R. Breu, A. Pretschner, 'Security Testing: A Survey', Advances in Computers, pp. 101, 1–51, 2016.

[15] M. Utting, B. Legeard, 'Practical model-based testing: a tools approach', 2010.

[16] B. Legeard, A. Bouzy, 'Smartesting certifyit: Model-based testing for enterprise it', Software Testing, Verification and Validation (ICST), IEEE Sixth International, pp. 391–397, March, 2013.

[17] J. Botella, F. Bouquet, J. Capuron, F. Lebeau, B. Legeard, F. Schadle, 'Model-Based Testing of Cryptographic Components', Software Testing, Verification and Validation (ICST), IEEE Sixth International Conference, pp. 192–201, March, 2013.

[18] J. Warmer, A. Kleppe, 'The object constraint language: Precise modeling with UML', Addison-Wesley Object Technology Series, 1998.

[19] C. Willcock, T. Deiß, S. Tobies, F. Engler, S. Schulz, 'An introduction to TTCN-3', John Wiley & Sons, 2011.

[20] G. Baldini, A. Skarmeta, E. Fourneret, R. Neisse, B. Legeard, F. Le Gall, 'Security certification and labelling in Internet of Things', IEEE 3rd World Forum on Internet of Things (WF-IoT), 2016.

[21] S. Pérez, J.A. Martínez, A. Skarmeta, M. Mateus, B. Almeida, P. Maló, "ARMOUR: Large-Scale Experiments for IoT Security & Trust", IEEE 3rd World Forum on Internet of Things (WF-IoT), 2016.

[22] T. Rakotoarivelo, M. Ott, G. Jourjon, I. Seskar, 'OMF: a control and management framework for networking testbeds', ACM SIGOPS Operating Systems Review, Vol 4, pp. 54–59, 2010.

[23] ARMOUR, 'Deliverable D4.1 – Definition of the large-scale IoT Security & Trust benchmarking methodology', 2016.

[24] Security Regulation, Conformity Assessment & Certification Final Report – Volume I: Main Report, Brussels October 2011. http://ec.europa.eu/dgs/home-affairs/e-library/documents/policies/security/pdf/secerca_final_report_volume__1_main_report_en.pdf. Last accessed 4/June/2017.

[25] Proposals from the ERNCIP Thematic Group, "Case Studies for the Cyber security of Industrial Automation and Control Systems", for a European IACS Components Cyber security Compliance and Certification Scheme, https://erncip-project.jrc.ec.europa.eu/component/jdown loads/send/16-case-studies-for-industrial-automation-and-control-sys tems/60-proposals-from-the-erncip-thematic-group-case-studies-for-the-cyber-security-of-industrial-automation-and-control-systems-for-a-european-iacs-components-cyber-security-compliance-and-certifica tion-scheme?option=com_jdownloads. Last accessed 4/June/2017.

[26] Minutes of the Joint EC/ENISA SOG-IS and ICT certification workshop. October 2014 ENISA. https://www.enisa.europa.eu/events/sog-is/minutes. Last accessed 4/June/2017.

[27] Kaluvuri, S. P., Bezzi, M., & Roudier, Y. (2014, September). A quantitative analysis of common criteria certification practice. In International Conference on Trust, Privacy and Security in Digital Business (pp. 132–143). Springer International Publishing.

[28] Antoine Coutant. French Scheme CSPN to CC evaluation http://www.yourcreativesolutions.nl/ICCC13/p/CC%20and%20New%20Techniques/Antoine%20COUTANT%20-%20CSPN%20to%20CC%20Evaluation.pdf. Last accessed 4/June/2017.

[29] "Common Criteria Reforms: Better Security Products through Increased Cooperation with Industry", available at: http://www.niap-ccevs.org/cc_docs/CC_Community_Paper_10_Jan_2011.pdf. Last accessed 4/June/2017.

[30] Anderson, R., & Fuloria, S. Certification and evaluation: A security economics perspective. In Emerging Technologies & Factory Automation, 2009. ETFA 2009. IEEE Conference on (pp. 1–7). IEEE. September 2009.

[31] General Data Protection Regulation. Regulation (EU) 2016/679 of the European Parliament and of the Council of 27 April 2016 on the protection of natural persons with regard to the processing of personal data and on the free movement of such data, and repealing Directive 95/46/EC (General Data Protection Regulation) http://eur-lex.europa.eu/legal-content/EN/TXT/?uri=CELEX:32016R0679. Last accessed 4/June/2017.

[32] EU privacy seals project. Inventory and analysis of privacy certification schemes. Rowena Rodrigues, David Barnard-Wills, David Wright. http://bookshop.europa.eu/en/eu-privacy-seals-project-pbLBNA26190/ ISBN: 978-92-79-33275-3. Last accessed 4/June/2017.

[33] TRUSTe Privacy Certification Standards, https://www.truste.com/privacy-certification-standards/ Last accessed 4/June/2017.

[34] OWASP, https://www.owasp.org/index.php/OWASP_Internet_of_Things_Project

[35] GSMA Security Framework CLP11, February 2016.

[36] NIST, CSRC, http://csrc.nist.gov/groups/SMA/fisma/framework.html

[37] "OneM2M security solutions", oneM2M-TR-0008-Security-V1.0.0, 2014.

8

IoT European Large-Scale Pilots – Integration, Experimentation and Testing

Sergio Guillén[1], Pilar Sala[1], Giuseppe Fico[2], María Teresa Arredondo[2], Alicia Cano[3], Jorge Posada [3], Germán Gutiérrez[3], Carlos Palau[19], Konstantinos Votis[20], Cor Verdouw[4,5], Sjaak Wolfert[4,5], George Beers[4], Harald Sundmaeker[6], Grigoris Chatzikostas[7], Sébastien Ziegler[8], Christopher Hemmens[8], Marita Holst[9], Anna Ståhlbröst[9], Lucio Scudiero[10], Cesco Reale[10], Srdjan Krco[11], Dejan Drajic[11], Markus Eisenhauer[12], Marco Jahn[12], Javier Valiño[13], Alex Gluhak[14], Martin Brynskov[15], Ovidiu Vermesan[16], François Fischer[17] and Olivier Lenz[18]

[1]MYSPHERA, Spain
[2]Universidad Politécnica de Madrid, Spain
[3]MEDTRONIC IBERICA, Spain
[4]Wageningen Economic Research, Wageningen University & Research, The Netherlands
[5]Information Technology Group, Wageningen University & Research, The Netherlands
[6]ATB Bremen, Germany
[7]BioSense Institute, Serbia
[8]Mandat International, Switzerland
[9]Luleå University of Technology, Sweden
[10]Archimede Solutions, Switzerland
[11]DunavNET, Serbia
[12]Fraunhofer Institute for Applied Information Technology, Germany
[13]Atos Spain, Spain
[14]Digital Catapult, UK
[15]Aarhus University, Denmark
[16]SINTEF, Norway
[17]ERTICO, Belgium
[18]Federation Internationale de l'Automobile, Region I, Belgium
[19]Universidad Politécnica de Valencia, Spain
[20]Centre for Research & Technology – Hellas, Greece

Abstract

The IoT European Large-Scale Pilots Programme includes the innovation consortia that are collaborating to foster the deployment of IoT solutions in Europe through the integration of advanced IoT technologies across the value chain, demonstration of multiple IoT applications at scale and in a usage context, and as close as possible to operational conditions.

The programme projects are targeted, goal-driven initiatives that propose IoT approaches to specific real-life industrial/societal challenges. They are autonomous entities that involve stakeholders from the supply side to the demand side, and contain all the technological and innovation elements, the tasks related to the use, application and deployment as well as the development, testing and integration activities.

This chapter describes the IoT Large Scale Pilot Programme initiative together with all involved actors. These actors include the coordination and support actions CREATE-IoT and U4IoT, being them drivers of the programme, and all five IoT Large-Scale Pilot projects, namely ACTIVAGE, IoF2020, MONICA, SynchroniCity and AUTOPILOT.

8.1 IoT European Large-Scale Pilots Programme

The scope of the IoT European Large-Scale Pilots Programme is to foster the deployment of IoT solutions in Europe through the integration of advanced IoT technologies across the value chain, demonstration of multiple IoT applications at scale and in a usage context, and as close as possible to operational conditions. Specific pilot considerations include:

- Mapping of pilot architecture approaches with validated IoT reference architectures such as IoT-A enabling interoperability across use cases.
- Contribution to strategic activity groups that were defined during the LSP kick-off meeting to foster coherent implementation of the different IoT Large-Scale Pilots.
- Contribution to clustering their results of horizontal nature (interoperability approach, standards, security and privacy approaches, business validation and sustainability, methodologies, metrics, etc.).

The IoT European Large-Scale Pilots Programme includes projects addressing the IoT applications based on European relevance, technology readiness and socio-economic interest in Europe. The IoT Large-Scale Pilot projects overview is illustrated in Figure 8.1, and the areas addressed by the projects are listed below.

Figure 8.1 IoT European Large-Scale Pilots Programme Projects Overview.

Research and innovation effort in specific IoT topics ensure the longer-term evolution of Internet of Things and the IoT European Large-Scale Pilots Programme projects are addressing:

- The integration and further research and development, where needed, of the most advanced technologies across the value chain (components, devices, networks, middleware, service platforms, application functions) and their operation at large scale to respond to real needs of end-users (public authorities, citizens and business), based on underlying open technologies and architectures that may be reused across multiple use cases and enable interoperability across those.
- The validation of user acceptability by addressing, in particular, issues of trust, attention, security and privacy through pre-defined privacy and security impact assessments, liability and coverage of user needs in the specific real-life scenarios of the pilot.
- The validation of the related business models to guarantee the sustainability of the approach beyond the project.

The IoT Large-Scale Pilots make use of the rich portfolio of technologies and tools so far developed and demonstrated in reduced and controlled environments and extend them to real-life use case scenarios with the goal of validating advanced IoT solutions across complete value chains with actual users and proving its socio-economic potential.

Support actions provide consistency and linkages between the pilots and complement them by addressing horizontal challenges critically important for the take-up of IoT at the anticipated scale. These include ethics and privacy, trust and security, respect for the scarcity and vulnerability of human attention, validation and certification, standards and interoperability, user acceptability and control, liability and sustainability.

The projects together form the IoT European Large-Scale Pilots Programme and a coordination body ensures an efficient interplay of the various elements of the IoT Focus Area and liaises with relevant initiatives at EU, Member States and international levels. The coordination is implemented by the creation of Activity Groups that are addressing topics of common interest across the Large-Scale Pilots.

8.2 ACTIVAGE – Activating Innovative IoT Smart Living Environments for Ageing Well

ACTIVAGE is a European Multi Centric Large-Scale Pilot on Smart Living Environments. The main objective is to build the first European IoT ecosystem across 9 Deployment Sites (DS) in seven European countries (see Figure 8.2), reusing and scaling up underlying open and proprietary IoT platforms, technologies and standards, and integrating new interfaces needed to provide interoperability across these heterogeneous platforms. This ecosystem will enable the deployment and operation at large scale of Active and Healthy Ageing IoT based solutions and services, supporting and extending the independent living of older adults in their environments, and responding to real needs of caregivers, service providers and public authorities.

8.2.1 Introduction

Throughout Europe and all around the world, mortality rates have fallen significantly over the past decades [1] leading to considerable changes in the age distribution of societies [2, 3]. In this context, people aged 60 are now expected to survive an additional 18.5 to 21.6 years and soon the world will have a higher number of older adults than children. This transformation is expected to continue, with the age group of elders (65+) growing from 18% to 28% of the EU population by the year 2060. Furthermore, according to the 2015 Ageing Report [4], one in three Europeans will be over 65 with a ratio of "working" to "inactive" population of 2 to 1, this representing a heavy impact on health and social care systems. Indeed, population ageing creates a common challenge for European countries as they must find ways to do more with less. Citizen empowerment and incitation to self-equip is one of the explored options.

Figure 8.2 ACTIVAGE Deployment Sites in 7 EU countries.

All of the aforementioned findings, highlighted the Active ageing and independent living activities (such as the EIP Action Group C2 – action plan) as the final goals for EU initiatives related to older adults [5].

Thus, legislation, technology, and reimbursement charges, enforce the health and social care systems to improve the way they are providing services to the European citizens. The "Active and Healthy Ageing" (AHA) community is wide and heterogeneous in terms of needs, demands and living environments [6]. AHA services based on the Internet-of-Things are promising to be a strategic component to support the creation of ecosystem able to dynamically answer and prevent the challenges faced by the health and social care systems [7] (H&SCS): the "always-connected" paradigm is becoming a way of life, and this could result in a positive transformation for H&SCS who are looking for new ways to reorient the provision of care and keep older people active and independent for longer.

8.2.2 Project Description

8.2.2.1 Main concepts in ACTIVAGE

ACTIVAGE is designed as ONE multi-centric Large-Scale Pilot across Europe. ACTIVAGE brings UNITY of objectives, evaluation methodologies, co-operation to achieve critical mass, and a single European platform to create and share evidence. Deployment Sites (DS) join clusters of stakeholders in the Active and Healthy Living value network, working together within a geographical space (a city or a region). These clusters or AHA-Business Ecosystems are mainly composed by a cohort of users (older adults, formal and informal caregivers), service providers; health care/social care administration; technological infrastructures and technology providers (infrastructure, sensors, applications, etc.).

DSs will deploy Reference Use Cases (UC) (see Figure 8.3) that address specific end-user needs, to improve their quality of life and autonomy. A single common interoperable ACTIVAGE IoT Ecosystem Suite (AIOTES) will be built up that provides every DS with the capacity to develop standard and interoperable IoT ecosystems on top of legacy IoT platforms, or communication and data management infrastructures. GLOCAL Evaluation Framework (Local KPIs and global KPIs) will be designed and implemented to demonstrate and evaluate health & social outcomes and socio-economic

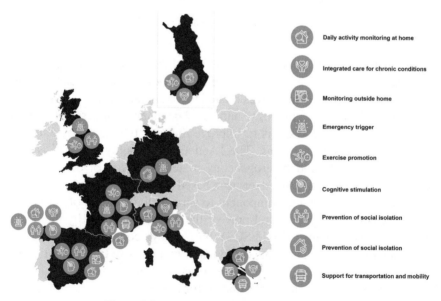

Figure 8.3   ACTIVAGE uses cases distribution.

impact from local up to a European scale, enabling effective exchange of experiences and cooperation among peers (e.g. users, providers, policy makers). 9 DS rolled out in 7 countries to constitute a major breakthrough to sustain open innovation in AHA field.

8.2.2.2 Targeted users and user needs

Within ACTIVAGE, "Ageing Well with IoT" is considered as the goal to extend healthy living years of older adults living independently and autonomously in their preferred environments by the massive adoption of IoT solutions. One of the most accepted measurement scales in different studies in Europe and World Wide is the Clinical Frailty Scale [8]: ACTIVAGE will concentrate on IoT solutions for older people classified under categories 1 to 5: "Very fit", "Well, Managing well", "Vulnerable" and "Mildly frail". ACTIVAGE will focus on deploying IoT solutions that work towards keeping older people away from category 6 and beyond, which already represents a significant cost in care for informal carers and for the formal healthcare systems.

ACTIVAGE focusses on "domains of needs" for the support of the older population and in order to create a demand-driven experience on the basis of

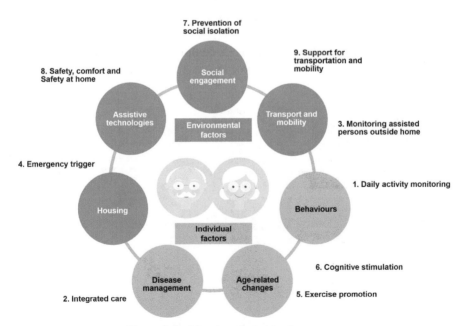

Figure 8.4   Mapping of needs and use cases.

the reference Ageing Well initiatives around the world. Figure 8.4 shows the domains of needs (on circles) and the Use Cases (UC) that will be deployed in the 9 DS involving up to 7,000 users.

8.2.3 The ACTIVAGE Model of IoT Ecosystem for Active and Healthy Ageing

ACTIVAGE is committed to build the first European AHA-IoT Ecosystem which is modelled as a technological infrastructure of hardware-software-services and standard protocols i.e. the "ACTIVAGE AIOTES", and a constellation of stakeholders interacting with each other within a governance framework towards the achievement of common goals.

Figure 8.5 shows this conceptual scheme. Data is the core asset of the ecosystem. Private Data is produced by wearable and medical devices and smart sensors and devices in their living environments (e.g. home, car, public spaces, etc.). Public data might come from different sources, not necessary linked to user interactions (e.g. weather, public services time tables, etc.). Personal/private data might be processed at the edge and at cloud level. If appropiate, interoperability interfaces are provided, enabling the delivery of a huge amount of data to be channeled across the pipeline and eventually feed hundreds or thousands of services for senior people, providers and payors.

The realization of this model is accomplished in ACIVAGE by the AIOTES: this is comprised of two layers that form all the necessary

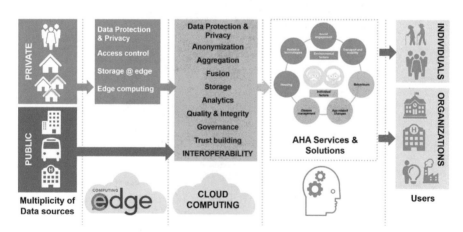

Figure 8.5   Model of AHA-IoT ecosystem.

components for: a) the support of a universal interoperability framework for the integration of the widest possible spectrum of platforms in the area of Active and Health Living and b) the formation of the diverse application marketplace and a set of application tools for the support of creators in the development and deployment phases of new applications. Figure 8.6 summarizes the envisioned architecture. The final architecture will be in compliance with standardisation projects such as IoT-A. The ACTIVAGE system will be separated into two distinct layers.

The IoT Interoperability Layer is aiming to efficiently and effectively integrate a wide spectrum of open and commercial platforms and IoT devices, having as a starting point the platforms provided for the ACTIVAGE and the IoT devices used in the Deployment Sites as shown in Figure 8.6. This layer will be further separated into two frameworks that will create standardized interfaces for a) the sharing of data with sensors and devices and b) the interoperability of their offered services. The Services Layer will include a number of functionalities to support efficient integration and effective deployment of new services to the envisioned ecosystem: the Applications Support Tools and the Marketplace.

8.2.4 Expected Project Impacts

The ACTIVAGE project has established a set of strategic impacts aligned with the vision of the project and designed to drive the activities across the project:

- **Societal impact:** ACTIVAGE may create evidence that support how AHA services based on IoT improve the quality of life of older adults, supporting the long-term sustainability and efficiency of health and social care systems. Aligned with this impact, ACTIVAGE may give answer to users' empowerment with the control of their data, safety and wellness, promoting healthy and active ageing, while enhancing the competitiveness of EU industry through new business models and expansion in new markets.
- **Innovation impact:** ACTIVAGE may ignite the economic growth by influencing the strategic decisions of the different stakeholders, offering a value based proposal built by the co-creation of the different stakeholders' views and interest. Only a valuable proposal with an integral multi-stakeholders commitment will assure the following key aspects: a) public and private investment by health and social care makers,

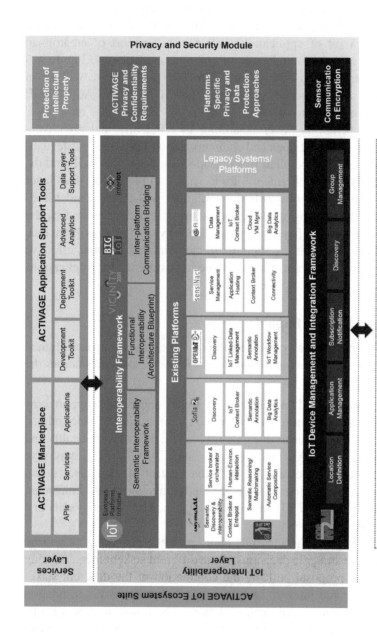

Figure 8.6   Conceptual architecture of AIOTES.

b) service providers adoption of cost-effective business models, c) senior citizens involvement in the creation and adoption of AHA services, d) to foster industrial innovation on IoT, wearables and sensor technologies and support standards for interoperability, and e) creation of new reliable and useful AHA solutions and services by SMEs and entrepreneurs.

- **Economic impact:** ACTIVAGE project may analyse the potential saving and contributions to the sustainability of the health care and social systems, and the readiness and maturity of the IoT technologies and ecosystem to host local solutions and to import and replicate solutions from other providers/location focused on value-based criteria.

8.2.5 Summary

During its 42 months duration the project will be aligned along a single five phase innovation path. After evaluating and demonstrating evidence and value to stakeholders in local Use Cases (UC), DSs will cooperate bilaterally or tri-laterally, allowing coherent, complementary replication of UC, to generate evidence on the value of interoperability and standardisation at a European scale. In the last phase, DSs will open to European external actors to incubate new UCs, technologies, solutions and Business Cases. Open calls will attract entrepreneurs and start-ups to implement innovative solutions using the mature DS's IoT ecosystem for testing, demonstration and initial market take-up.

8.3 IoF2020 – Internet of Food and Farm 2020

The Internet of Things (IoT) is expected to be a real game changer that will drastically improve productivity and sustainability in food and farming. However, current IoT applications in this domain are still fragmentary and mainly used by a small group of early adopters. The Internet of Food and Farm 2020 Large-Scale Pilot (IoF2020) addresses the organizational and technological challenges to overcome this situation by fostering a large-scale uptake of IoT in the European food and farming domain. The heart of the project is formed by a balanced set of multi-actor trials that reflect the diversity of the food and farming domain. Each trial is composed of well-delineated use cases developing IoT solutions for the most relevant challenges of the concerned subsector. The project conducts 5 trials with a total of 19 use cases in arable, dairy, fruits, vegetables and meat production. IoF2020 embraces a lean multi-actor approach that combines the development of Minimal Viable Products (MVPs) in short iterations with the active involvement of various stakeholders. The architectural approach supports interoperability of multiple use case systems and reuse of IoT components across them. Use cases are also supported in developing business and solving governance issues. The IoF2020 ecosystem and collaboration space is established to boost the uptake of IoT in Food and Farming and pave the way for new innovations.

8.3.1 Introduction

IoT is a powerful driver that is expected to transform the entire farming and food domain into smart webs of connected objects that are context-sensitive and can be identified, sensed and controlled remotely [9–11]. IoT will be a real game changer in agriculture that drastically improves productivity and sustainability. This vision is illustrated by the story in Figure 8.7, which is an

---

March 2020, a field somewhere in Europe

The morning mist soaks into thick shreds across the country, above the sun rises and turns the horizon red. From the fog a soft humming sound, two tractors emerge. When he spots me, the driver of the second tractor steps out, but where is the driver of the first tractor? There is none, says the farmer, I operate both machines. How? Well, that strange vehicle you saw here last week has mapped the whole field and this map is now instructing the board computers of the two tractors how to drive. The first tractor exactly follows pre-programmed lines and carries out soil cultivation, based on soil composition. My tractor with a sowing machine automatically follows the same lines and automatically adjusts distance, quantity and variety of potato seed. Incredible, isn't it?

Two weeks later...

The same field. An unmanned small tractor drives with a high speed along the same invisible straight lines. With surgical precision, a hoe eliminates every weed in the field, the farmer says. This saves a lot of chemicals and labour in comparison to earlier days where we had to spray the full field with a heavy tractor. So this is good for the environment and I have much less costs! Within a few weeks the fertilizer will follow the same lines again and by a pre-defined task map it knows exactly where and what to put different types of fertilizer for optimal growth of the plants. That map was generated on the basis of big data analysis and calculations in the cloud involving relevant data from the market, weather and public regulations. Additional cameras are checking the crop and, if necessary, make corrections. Again, the plants just get enough nutrients to grow optimally and nothing is spoiled to the environment. Wow, amazing! Come, I'll show you how it works in the office. Don't you have to stay with your tractor? Oh no, it knows what it is doing.

At the office with a good old-fashioned cup of coffee...

Of course we farmers are still in charge of our own farm but most of the field operations are carried out automatically by autonomous objects. Now we can focus on the market choices and take care of communication with our customers and last but not least citizens who are very much involved in farming nowadays. After execution of the field work, the measured data is automatically returned from the machine to the office through the cloud. This is the basis for subsequent tasks. But I also provide it to research institutes, which feed these data into computer models for further improvement. The same holds for public legislation and certification bodies. They use the same data to check for compliance to their rules. Every organisation has access to a specific set of our data in the cloud. Of course, this is subject to strict security and privacy rules. No, no, I don't want leave my data lying around. Oh yes, by the way, food safety and traceability is not an issue anymore; it is highly guaranteed by all kind of sensors and in case something might go wrong early warning systems alert me in time.

Figure 8.7   Illustrative story of the vision on IoT in agriculture.

example for arable farming, but it is exemplary for other subsectors such as dairy, meat, vegetables, and fruits including wine and olives.

To make this vision come true much technology is already available, although there are specific IoT challenges in this sector. Agri-Food 'things' are often living objects and attached devices have to work in harsh environments, while network connectivity in rural areas can be challenging. In fact, a large-scale uptake of IoT in agriculture is in particular prevented by a lack of interoperability, user concerns about data ownership, privacy and security, and by appropriate business models that are also suitable for (very) small companies [12, 13]. Consequently, current IoT applications in farming and food are still fragmentary and mainly used by a small group of early adopters, despite the great world-wide interest of IoT technology providers and investors.

IoF2020 is a European Large-Scale Pilot (LSP) on IoT for Smart Farming and Food Security. Its main objective is to foster a large-scale uptake of IoT in the European farming and food domain. This will contribute to a next huge innovation boost and consequently to a drastically improved productivity and sustainability in the agri-food domain. More specifically, IoF2020 aims to:

- Demonstrate the business case of IoT for a large number of application areas in farming and food;
- Integrate and reuse available IoT technologies by exploiting open architectures and standards;
- Ensure user acceptability of IoT solutions in farming and food by addressing user needs, including security, privacy and trust;
- Ensure the sustainability of IoT solutions beyond the project by validating the related business models and setting up an IoT Ecosystem for large scale uptake.

The IoF2020 consortium consists of 71 public and private partners from 16 different countries and has a total budget of 35 M€. The project started in January 2017 and will last for 4 years.

8.3.2 Trials and Use cases

The heart of the project is formed by a balanced set of multi-actor trials that reflect the diversity of the food and farming domain, including different agricultural sub sectors, conventional and organic farming, early adopters and early majority farmers, SMEs and large industrial companies, and different supply chain roles including logistics and consumption. Each trial is composed of well-delineated use cases that together address the most relevant challenges for the concerned subsector. The use cases follow a demand-driven philosophy in which IoT solutions for specific business needs are developed by a dedicated team of agri-food end users and IoT companies (integrators, app/service developers, infrastructure/technology providers) with a clear commercial drive, supported by R&D organisations. IoF2020 conducts 5 trials with a total of 19 use cases in arable, dairy, fruits, vegetables and meat production (Figure 8.8).

 The ***Internet of Arable Farming (trial 1)*** integrates operations across the entire arable cropping cycle combining IoT technologies, data acquisition (soil, crop, climate) in growing and storage of arable crops (potatoes, wheat and soya beans). These will be linked to existing sensor networks, earth observation systems, crop growth models and yield gap analysis tools and external databases (e.g. economic/environmental impact) and translated into farm management systems. The trial will result in increasing yields, less environmental impact, easier cross-compliance and product traceability and more use of technology by farmers. The trial consists of 4 use cases:

1.1 *Within-field management zoning*: defining specific field management zones by developing and linking sensing- and actuating devices with external data;

Figure 8.8   Geographical coverage of the IoF2020 trials and use cases.

1.2 *Precision Crop Management*: smart wheat crop management by sensors data embedded in a low-power, long-range network infrastructure;

1.3 *Soya Protein Management*: improving protein production by combining sensor data and translate them into effective machine task operations;

1.4 *Farm Machine Interoperability*: data exchange between field machinery and farm management information systems for supporting cross-over pilot machine communication.

The ***Internet of Dairy Farming (trial 2)*** implements, experiences and demonstrates the use of real-time sensor data (e.g. neck collar) together with GPS location data to create value in the chain from 'grass to glass', resulting in more efficient use of resources and production of quality foods, combined with a better animal health, welfare and environment implementation. The trial focuses on feeding and reproduction of cows through early warning systems and quality data that can be used for remote calibration and validation of sensors and consists of 4 use cases:

2.1 *Grazing Cow Monitor*: monitoring and managing the outdoor grazing of cows by GPS tracking within ultra-narrow band communication networks;

2.2 *Happy Cow*: improving dairy farm productivity through 3D cow activity sensing and cloud machine learning technologies;

2.3 *Silent Herdsman*: herd alert management by a high node count distributed sensor network and a cloud-based platform for decision-making;

2.4 *Remote Milk Quality*: remote quality assurance of accurate instruments and analysis & pro-active control in the dairy chain.

The ***Internet of Fruits (trial 3)*** demonstrates IoT technology that is integrated throughout the whole supply chain from the field, logistics, processing to the retailer. Sensors in orchards and vineyards (incl. weather stations, multispectral/thermal cameras) will be connected through the cloud and used for monitoring, early warning of pests and diseases and control (e.g. variable rate spraying, selective harvesting). Traceability devices (incl. RFID, multidimensional barcodes) and smart packaging allows for condition monitoring during storage, processing, transportation and on the shelves. Big data analyses will further optimize all processes in the whole chain. This will result in reduced pre- and post-harvest losses, less inputs, higher (fresh) quality and better traceable products (incl. protected designation of origin, PDO). The trial consists of 4 coherent use cases:

3.1 *Fresh table grapes chain*: real-time monitoring and control of water supply and crop protection of table grapes and predicting shelf life;

3.2 *Big wine optimization*: optimizing cultivation and processing of wine by sensor-actuator networks and big data analysis within a cloud framework;

3.3 *Automated olive chain*: automated field control, product segmentation, processing and commercialisation of olives and olive oil;

3.4 *Intelligent fruit logistics*: fresh fruit logistics through virtualization of fruit products by intelligent trays within a low-power long-range network infrastructure.

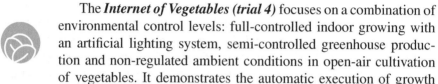

The ***Internet of Vegetables (trial 4)*** focuses on a combination of environmental control levels: full-controlled indoor growing with an artificial lighting system, semi-controlled greenhouse production and non-regulated ambient conditions in open-air cultivation of vegetables. It demonstrates the automatic execution of growth recipes by the intelligent combination of sensors that measure crop conditions and control processes (incl. lighting, climate, irrigation and logistics) and analysis of big data that is collected through these sensors and advanced visioning systems with location specification. This will result in improved production control and better communication throughout the supply chain (incl. harvest prediction, consumer information). The trial consists of 4 use cases:

4.1 *City farming*: value chain innovation for leafy vegetables in convenience foods by integrated indoor climate control and logistics;

4.2 *Chain-integrated greenhouse production*: integrating the value chain and quality innovation by developing a full sensor-actuator-based system in tomato greenhouses;

4.3 *Added value weeding data*: boosting the value chain by harvesting weeding data of organic vegetables obtained by advanced visioning systems;

4.4 *Enhanced quality certification system*: enhanced trust and simplification of quality certification systems by use of sensors, RFID tags and intelligent chain analyses.

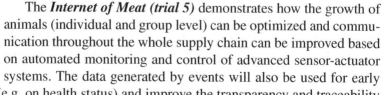

The ***Internet of Meat (trial 5)*** demonstrates how the growth of animals (individual and group level) can be optimized and communication throughout the whole supply chain can be improved based on automated monitoring and control of advanced sensor-actuator systems. The data generated by events will also be used for early warning (e.g. on health status) and improve the transparency and traceability of meat throughout the whole supply chain. This will assure meat quality, reduce mortality, optimize labour and improve animal health and welfare leading to reduction of antibiotics. The trial consists of 3 use cases:

5.1 *Pig farm management*: optimise pig production management by interoperable on-farm sensors and slaughter house data;

5.2 *Poultry chain management*: optimize production, transport and processing of poultry meat by automated ambient monitoring & control and data analyses;

5.3 *Meat Transparency and Traceability*: enhancing transparency and traceability of meat based on an monitored chain event data in an EPCIS-infrastructure.

8.3.3 Technical Architectural Approach

Each use case will be an autonomous implementation of an IoT system, which provides a dedicated solution for a specific domain challenge. However, for a large scale uptake it is important to maximize synergies across multiple use case systems. Therefore, a core concept of IoF2020 is that the use case systems function as nodes in a software ecosystem [14]. As a consequence, much attention is paid to ensuring the interoperability of multiple use case systems and the reuse of IoT components across them. Figure 8.9 shows the architectural approach to achieve this during design, development, implementation and deployment.

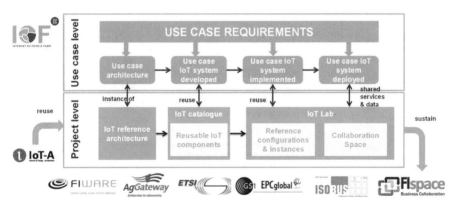

Figure 8.9 The IoF2020 architectural process to ensure reuse and interoperability of use case IoT systems.

The use case architectures will be based on a common technical reference architecture to create a shared understanding and to maximize synergies across multiple use case systems. Each use case within a trial will design a specific instance of the reference architecture to address its specific user requirements. The project will provide a catalogue of reusable system components, which can be integrated in the IoT systems of multiple use cases to facilitate large-scale uptake. This repository goes beyond a checklist and includes practical guidelines and implementation tools. The IoF2020 lab will support the implementation of reusable IoT components in a testbed environment. Finally, IoF2020 will provide a Collaboration Space in which services and data can be shared as a key enabler to facilitate the interaction between the IoT systems of the use cases during deployment. As indicated, the project will reuse components and knowledge from previous projects and existing organizations and try to embed and sustain the project results into the same organizations.

8.3.4 Lean Multi-Actor Approach

IoF2020 embraces a demand-driven methodology in which end-users from the agri-food are actively involved during the entire development process aiming at cross-fertilisation, co-creation and co-ownership of results (see Figure 8.10).

The approach for the use cases is a combination of the *lean start-up methodology* that focuses on the development of Minimal Viable Products

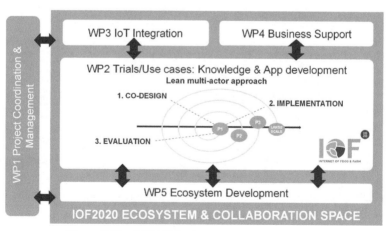

Figure 8.10   IoF2020 Project approach and structure.

(MVPs) in short iterations and the *multi-actor approach* that stresses the active involvement of various stakeholders. The use cases will actively be supported by three other work packages (WPs). WP3 facilitates sharing, reuse and finally integration of IoT components as described in the previous section. WP4 provides business support in terms of monitoring KPIs, business models, market studies and governance aspects (incl. security, data ownership, privacy, liability and ethical issues). WP5 facilitates the development and expansion of the various ecosystems on use case and project level and beyond amongst others by communication, dissemination, organizing workshops and events and by active involvement of European and national communities from the demand- and supply-side of IoT, including industry associations and cooperatives, European Innovation Partnerships, Technology Platforms and ERAnets. A mid-term open call of 6 M€ will be used to further accelerate these developments. This approach establishes a large IoF2020 ecosystem and collaboration space that is expected to sustain after the project.

8.3.5 Conclusion and Outlook

IoF2020 aims to boost the uptake of IoT in European Food and Farming. This will be realized through a balanced set of multi-actor trials and use cases in several subsectors. The use cases are developed in a scalable manner through an open technical architecture and infrastructure with components that can be shared and reused by stakeholders outside the project. This development is leveraged by activities that build-up and extend the total ecosystem, defining attractive and successful business models and solving governance issues. In this way IoF2020 will pave the way for data-driven farming, autonomous operations, virtual food chains and personalized nutrition for European citizens.

8.4 MONICA – Management of Networked IoT Wearables – Very Large Scale Demonstration of Cultural Societal Applications

The Large-Scale Pilot MONICA demonstrates how cities can use the Internet of Things to deal with sound, noise and security challenges at big, cultural, open-air events. A range of applications will be demonstrated in six major European cities involving more than 100,000 users in total. The project brings together 29 partners from 9 European countries with the objectives to provide

Figure 8.11   An example of the type of open air events that will be addressed by MONICA: here the Kappa Futur Festival in Turin, Italy. MONICA will try to improve the sound experience on events and at the same time reduce the noise for the neighbours. It will also improve responsiveness to security challenges[1].

a very large-scale demonstration of multiple existing and new Internet of Things technologies for Smarter Living.

Imagine sound zones at outdoor concerts in the city where the sound experience is enhanced for those who enjoy the music and the noise mitigated for those who don't. Visualise intelligent cameras deployed at city festivals which, while preserving privacy, estimate crowd size and density in real time, notifying security staff of any unusual crowd behaviour. Or imagine smart wristbands and mobile apps, allowing people to interact with each other and the performers, informing people of the best way out of the venue or guiding them to the nearest exit in case of an emergency.

These are some of the several applications which MONICA will demonstrate at minimum 16 cultural events, taking place all over Europe in Copenhagen, Bonn, Hamburg, Leeds, Lyon and Torino. The broad list of events includes concerts, festivals, city and sport events and involve the use of multiple, wearable, mobile and fixed devices with sensors, such as

---

[1] Photo courtesy of Simone Arena SIMPOL-lab.

Table 8.1 Overview of Pilot Cities and Events

	City	Event	Avg. Number of visitors per year
	Copenhagen (DK)	• Friday Rock at Tivoli	48.000
	Torino (IT)	• KappaFutur Festival • The Movida	18.000 80.000
	Hamburg (DE)	• Hamburger DOM • Port Anniversary	1.500.000 1.000.000
	Lyon (FR)	• Nuits Sonores • La Fête des Lumières	100.000 3.000.000
	Bonn (DE)	• Rhein in Flammen • Pützchens Markt	300.000 1.350.000
	Leeds (UK)	• Cricket & Rugby matches at Headingley Stadium)	48.000

wristbands, smart glasses, video cameras, loudspeakers, drones and mobile phones. The full list of pilot events is shown in Table 8.1.

8.4.1 Introduction

The deployment of Internet of Things has also a major impact on society specifically in urban environments, where it helps to solve major societal challenges. The rapidly growing number of Smart City platforms enables cities to assemble all their digital applications on uniform communication networks spanning entire cities delivering diverse applications such as health, energy and resource efficiency, and traffic management that help the city to become more environmentally sustainable and citizens to have a better life. In technology areas, standard IoT middleware, architecture and technology enablers have mitigated the complexity of communication and integration and paved the way for a wealth of innovative distributed applications in many vital areas of our society. However, most of the IoT and Smart City platforms are still insufficiently developed to handle really large scale deployment. Health and smart living may potentially involve thousands of users, but they are relatively scarcely distributed, even within a city, and the communication load is limited.

In response, the MONICA platform will demonstrate a resilient IoT platform that addresses major issues of large scale deployments: Scaling, costs of sensors, and intelligence. On this background, the MONICA project is uniquely innovative since it will demonstrate an extremely large uptake of a multitude of IoT applications (10,000+ simultaneous and more than 100,000 different users) using low-costs wearables and apps running on existing wearable platforms such as smart watches and smart phones in combination with wearable sensors. Moreover, the platform will demonstrate heterogeneous interfaces for both expensive, professional infrastructure components and affordable, wearable, and widely used consumer devices. Finally, the demonstration will show closed feedback loops to actuating networks and human interaction and intervention based on situational awareness and decision support.

8.4.2 The MONICA Ecosystems

The MONICA IoT Platform will be demonstrated in the scope of three ecosystems:

8.4.2.1 The Security Ecosystem

The MONICA Security Ecosystem will demonstrate how a multitude of innovative applications for managing public security and safety can be seamlessly integrated with IoT sensors and actuators and used in large scale. The core security and safety challenges at large events, such as those proposed in MONICA, are the handling and mitigation of unforeseen incidents and accidents: personal violence, panic scenes, severe illness of individuals in the middle of a crowd, infrastructure catastrophes such as fire or structural collapses. The aim of any security platform is to ensure the monitoring, recording, identification, analysis of any part of the monitored environment, and measures capable of predicting and, whenever possible, mitigate the danger of potential or imminent events. The modelling of incidents and accidents is therefore necessary in order to be able to deal with episodes while or before they unfold. The Security Ecosystem will consist of a series of large and small applications that, in combination, can be used to monitor and manage the security situation before, during and after an event. The main objectives of the applications are to demonstrate how the IoT platform will seamlessly support open and closed loop solutions that address real-life safety challenges. The traditional security tools used at an event are normally a variety of perimeter security as e.g. fences, supported by CCTV cameras, few entrances with guards and guards working around the area. MONICA will implement additional security and safety measures, e.g. by combining information on the fly for real-time operation and to support the work of security staff. MONICA hence will not change the already existing security concepts of events, rather it will complement extra valuable real-time information supporting the early detection of potentially critical situations.

8.4.2.2 The Acoustics Ecosystem

The MONICA Acoustic Ecosystem will demonstrate how a multitude of innovative applications for managing open air music performances in the public space can be seamlessly integrated with sensors and actuators using the MONICA platform. It will consist of a series of large and small applications that, in combination, can be used to monitor and manage the sound before, during and after a performance. The main objectives of the applications are to demonstrate how the IoT platform will seamlessly support open and closed loop solutions that address real-life environmental challenges e.g. noise in public spaces. E.g. at inner city open air concerts, sound fields will be optimised with respect for both the performers and the concert audience in terms of loudness, directionality and quality. The sound zone system and actuation layer of the MONICA platform allows for dynamic adjustment of the active sound field control loudspeakers, thus, improving the sound quality of visitors while at the same time reducing the noise for neighbours. Health monitoring of sound level exposure can be offered to concerned concertgoers. A cheap, wearable sound level meter in bracelets can be connected to a Smartphone app and continuously measure the cumulative sound dose. The user can thus seek less loud sections of the concert arena.

8.4.2.3 The Innovation Ecosystem

Communication to customers, crowds and citizens is improved by the use of mobile apps and IoT wristbands with value-adding features, enabling people to interact with and locate each other, informing visitors of the best place to park, the best way out or the bars with the shortest queue, and guiding participants to the nearest exit in case of an emergency. General data such as on sound levels are made accessible as open data on the hosting city's websites for citizen engagement and innovation. Applications for user involvement with artists at concerts or among citizens based on different kinds of wearables will be tested. Open APIs will be provided to foster new businesses and start-up solutions based on the MONICA IoT Platform.

8.4.3 User-Driven Pilots

The MONICA platform and its components will be demonstrated in six different pilot cities in different five Member States across Europe and as close as possible to real-life operational conditions under typical load and constraints associated with organising big events. From the pilots, the project will evaluate both qualitative and quantitative success measures towards established KPI's related to stakeholder satisfaction and the improved efficiency in handling of large scale events. In this framework, the following specific activities will take place in the selected pilot sites:

- Integration, deployment and operation of commercially available fixed and mobile devices and development of new wearable devices in the MONICA IoT and relevant Smart City infrastructures.
- Integration, deployment and operation of commercially available fixed and mobile actuating devices and the relevant software applications mentioned above.
- Development and deployment of fully automated closed-loop systems which uses sensing inputs from the IoT network layers to assist humans in monitoring, situational awareness and decision making and provide the resulting control regimes for the actuating IoT infrastructure.
- Validation of pre-defined impact Key Performance Indicators (KPIs) related to each pilot site along with the methodology and the relevant assessment procedures in order to obtain and disseminate qualitative and quantitative date for replication.
- Demonstration of the generic applicability and interoperability of experimental testbeds and open platforms in validation of IoT technologies and identification of where standards are missing and pre-normative activities are needed.
- Development and validation of new markets and business models aiming at involving all actors in the innovation value chain as well as assessing the impact on Europe's Cultural and Creative Industries.
- Active involvement of all actors in the validation and dissemination in order to establish the best possible foundation for creating maximum impact and replication potential from the demonstrations.

In addition to the planned demonstrations, the MONICA deployments addresses – in a smaller scale – the solutions to more generic Smart City challenges i.e. the deployment of a common ICT infrastructure and components in city areas with very diverse needs and context, to deliver services with different business and technical configurations. The specification for the

Figure 8.12   Hamburg DOM is Northern Germany's biggest goose fair. It takes place three times a year in spring, summer and winter and offers its attractions to 7–10 million visitors during the 91 DOM days[2].

MONICA pilots will be extracted from comprehensive use cases and concrete business cases defined by demand-side stakeholders and users. Demand-side representatives in the requirements process will be drawn from the fields of concert organisers, artistic performers, spectators, public authorities, citizens, civic engagement groups, and other relevant groups found inside and outside the consortium. This methodology allows for maximum stakeholder input captured in the analysis of the use cases, the business ecosystem, the value chain interactions and the general societal and environmental realm.

The implementation and deployment of the pilot sides will be based in a mix of commercially available components and solutions, open architectures and design approaches from previous the portfolio of technologies and tools so far developed and demonstrated in reduced and controlled environments as well as targeted research and development of specific prototype solutions where needed. Pilot work plans will include feedback mechanisms to allow adaptation and optimisation of the technological and business approach to the particular use case.

---

[2]Photo courtesy of Hamburg.de

8.4.4 The MONICA Technical Concept

The MONICA platform features a cloud based on advanced, open IoT technologies dynamically integrating fixed and nomadic devices and truly mobile wearables in the physical world with automated closed-loop actuating functions. The platform will also integrate humans in the loop where appropriate, by providing situational awareness and dynamic decision support tools. A strong toolbox for security and trust management will complement the platform.

The platform will be able to support multiple IoT applications in a wide usage context focusing on the two most important challenges for organisers of large scale concerts and cultural events in large cities: Unwanted noise in the surroundings and security of the audience.

8.4.4.1 The MONICA architecture

The MONICA platform is built on several IoT physical world network infrastructures and a closed loop control system for each application. The components are connected via dedicated communication network and data repositories. The entire solution is embedded in a MONICA Private Cloud structure as visualised in Figure 8.13.

8.4.4.2 The MONICA IoT Infrastructure

The MONICA IoT Infrastructure, depicted in Figure 8.14, must be capable of handling three different types of IoT devices: i) wearable devices, ii) nomadic devices and iii) fixed sensors and fixed Cyber Physical Systems.

Wearable devices include wristbands, glasses and mobile phones. Wristbands are intended to be worn by the spectators and staff while glasses are intended mainly for the security staff. Wristbands have connectivity based on either Ultra-Wideband (UWB) or narrow-band radio (868/900 MHz) technologies. Smart glasses instead will be used, based on the Android OS and equipped with front-facing camera, inertial sensors, light sensor, GPS and pressure sensor. The glasses have real see-through displays with a WVGA resolution and have Wi-Fi b/g/n and BT4.0 connectivity. Nomadic devices are mobile devices confined to the event area, such as hand held sound dosimeters and other sensors (e.g. sound meter, temperature sensor, wind sensor, camera, etc.) mounted on controllable airships. Fixed sensors and Cyber Physical Systems comprise devices mounted on fixed structures in and around the event area, e.g. sound pressure gauges and dosimeters, microphones, cameras, anemometers, etc.

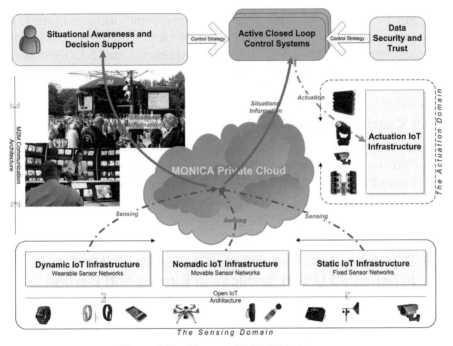

Figure 8.13 The overall MONICA concept.

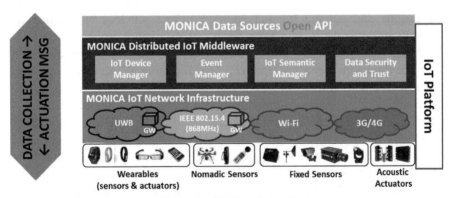

Figure 8.14 MONICA IoT Architecture.

In order to manage the heterogeneity of a large amount of the above-mentioned devices, a proper IoT architecture will be defined following the AIOTI High-level Architecture (HLA) and the AIOTI Domain Model proposed by the AIOTI WG03 – IoT Standardisation. The MONICA Distributed

IoT Middleware will implement the IoT infrastructure using existing IoT gateways and services, e.g. LinkSmart®, FI-WARE, SCRAL, and oneM2M. De-facto standards such as MQTT for publish/subscribe and SAREF (Smart Appliance Reference Ontology, ETSI TS 103 264 standard) and the W3C Semantic Sensor Network Ontology for semantic modelling will be applied.

8.4.5 Conclusion and Outlook

The MONICA project will carry out unique demonstrations of large scale take up of IoT deployments at highly relevant inner city cultural open-air events. It will pave the way for innovative business opportunities for technology and software providers in the field of IoT. Ultimately, the project aims to improve the quality of life in our cities for all citizens.

8.5 SynchroniCity: Delivering a Digital Single Market for IoT-enabled Urban Services in Europe and Beyond

Smart cities hold the potential to be a key driver and catalyst in creating a large scale global IoT market of services and hardware. However, the emerging smart city market faces specific challenges that act as barriers to growth, impeding rapid innovation and inhibiting widespread market adoption.

SynchroniCity is an ambitious initiative to deliver a **digital single market** for Europe and beyond **for IoT-enabled urban services** by piloting its foundations at scale in reference zones across eight European cities and involving other cities globally. It addresses how to incentivize and build trust for companies and citizens to actively participate and find common co-created IoT solutions for cities that meet citizen needs, and to create an environment of evidence-based solutions that can easily be replicated in other regions. These reference zones are based on cities at the forefront of smart city development covering different geographies, cultures and sizes and include Antwerp (BE), Carouge (CH), Eindhoven (NL), Helsinki (FI), Manchester (UK), Milano (IT), Porto (PT) and Santander (ES). Globally, SynchroniCity adds committed replicating reference zones in Mexico, Korea, USA and Brazil.

8.5.1 Introduction

Digital technologies offer an opportunity to profoundly change how our existing society works. They can enable a transformation of different industry

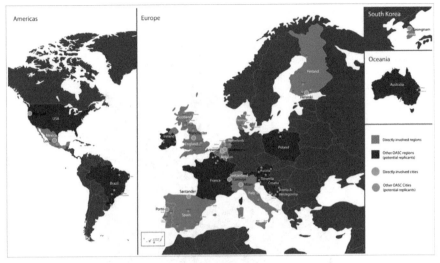

Figure 8.15   SynchroniCity cities and regions.

sectors improving existing business activities, processes, and competencies within organizations and across their boundaries.

Data infrastructures and the Internet of Things (IoT) form a critical part of the digital transformation of cities and communities by creating adequate awareness of real-world processes in order to drive more efficient, partially autonomous, decision making, while still maintaining a high level of data protection, inclusivity and general support for local priorities such as economic development and cultural heritage.

In terms of data infrastructures, cities have been at the forefront of embracing the open data movement. The release of data sets to the public has increased transparency and provided early innovation potential for third party stakeholders. Services such as Citymapper[3] show how open data can add great benefits to the journey experiences of citizens.

Many cities have invested in the setup of open data portals and proactively encourage stakeholders across public departments and the private sector to contribute data sets. At the same time, cities are trying to engage entrepreneurs and communities to innovate around these data stores. Early results are promising, but static or sporadically changing data sets have their limitations.

---

[3] https://citymapper.com/

Figure 8.16 Street light in Santander.

IoT infrastructures are increasingly becoming an important element in providing the underpinning digital layer of smart city services. They augment the open data sets with rich real-time information about public infrastructure conditions and city processes that can be exploited for a more responsive delivery of public services. Examples range from an improved mobility experience through adaptive traffic management and multi-modal transportation to resource savings achieved by smart street light control, waste collection and irrigation management.

Various demonstrations of such systems are emerging globally showing the benefits of data-driven services based on IoT and data infrastructures. However, many of these systems currently operate in silos both in terms of the technology employed and the operating environments of the city. Interoperability issues and lack of economies of scale make many potential business cases still hard to justify and result in a lack of confidence in the market.

SynchroniCity aims to overcome the existing barriers in the market by fostering the emergence of a digital single market for smart city services. It brings eight European cities together to work on a common blueprint for IoT and data infrastructures with standardized interfaces and information models, creating an environment that allows vendors and solution providers to more openly compete.

Our vision is to move from disparate data stores and city platforms to vibrant marketplaces for urban data and services providing adequate incentives for a variety of stakeholders to participate. For providers of IoT infrastructure and other urban data sources, this should provide a trusted environment to generate reliable revenue flows. For application and service developers, it should allow frictionless access to reliable and trusted urban data streams to be used as assets underpinning the innovation no matter what city is involved. We call this aspect "avoiding city lock-in". Cities and infrastructure providers can benefit from an aligned environment with standardised interfaces to access a diverse pool of vendor solutions able to compete fairly on price and performance. We call this aspect "avoiding vendor lock-in". Together, they form the robust underpinnings of a global market for IoT-enabled urban services.

8.5.2 Technical and Non-Technical Barriers of Creating a Smart City Eco-System

SynchroniCity addresses a wide range of technological and non-technological barriers that need to be overcome to enable necessary economies of scale and market confidence to emerge.

The key technological barriers include:

- Lack of standardized multi-vendor ecosystem, leading to fear of vendor lock-in for many cities;
- Lack of common service provisioning environments across cities, leading to fear of city lock-in for service developers as they need to "redevelop" major parts of their apps and renegotiate access to different data sources for every city;
- Close coupling of IoT infrastructure and applications, leading to IoT solution silos and limited infrastructure reuse;
- Lack of tools, license models and platforms to facilitate the incentivized sharing of urban IoT data and other relevant data sets;
- Lack of harmonized business practice and legal frameworks across cities – making IoT infrastructure roll out and launch of new services tedious across different geographies;

- Lack of understanding of privacy and personal data protection implications – making it difficult to fully leverage data of citizens collected in public spaces in responsible way; and
- Lack of confidence in adopting emerging technologies due to increasing technology fluidity – rapid change and emergence of new standards make it hard for cities to understand where sustainable investments can be made.

Besides technological challenges, cities offer unique constraints and non-technological barriers, which hamper the adoption of IoT technologies. They include:

- Economical costs and budget constraints make it difficult to for cities to make major investments in newly emerging technologies;
- Inflexible public tender processes and procurement models complicate experimental exploration with new technology solutions;
- Frequent political changes and reliance on election cycles leads often to reprioritisation and lack of continuity to focus on longer term innovation programmes;
- Lack of a holistic smart city strategy leads to investments in fragmented systems for different verticals, making it difficult to capitalise across these and gain cost savings; and
- Lack of involvement of citizens and support from these can lead to smart city solutions not addressing real citizen needs and fuel the mistrust of citizen in adopting new technology solutions.

8.5.3 SynchroniCity Technical Approach

Overcoming the barriers identified above requires a common approach across the different reference zones. In the following we introduce the key foundations for our vision of the SynchroniCity digital single market.

Technical barriers 1–3 and in part 7 demand a common reference architecture for smart city platforms. A standardised reference architecture, which is widely adopted among many cities with clearly defined components and interfaces is fundamental to overcome vendor lock-in. It will boost market confidence and lay down the foundations for the required economies of scale.

Key elements in this reference architecture are common north- and southbound interfaces. Technical barrier 4 demands new market place enablers that encourage sharing of urban IoT data and other relevant data sets among

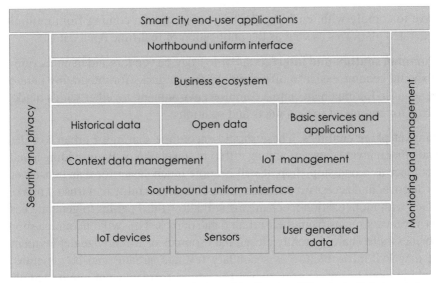

Figure 8.17 High level architectural view of the SynchroniCity single digital market.

different stakeholders. Lastly, barriers 5–6 relate to finding agreement on common principles of governance of a digital single market.

Figure 8.17 shows an overview of the proposed components of the SynchroniCity single market place including common north- and southbound interfaces. A further important feature is the market place enablers that underpin a thriving business eco-system around IoT data streams, actuation capabilities and other urban data sources. In the following we briefly describe each of these.

A common reference architecture for smart city platforms. A standardized reference architecture, which is widely adopted among cities with clearly defined components and interfaces, is fundamental to overcome vendor lock-in. It will boost market confidence and lay down the foundations for the required economies of scale.

Common northbound interfaces. Developers require a common, homogeneous and IoT independent way to access data from the devices infrastructure, but also from any other subsystem in the city that can provide valuable information to develop smart services and applications. More specifically, this includes 1) a common standard API for context information management; 2) a common set of information models enabling actual interoperability of applications; and 3) a set of common standards data publication platforms have to comply with, enabling the harvesting of data coming from multiple federated platforms as well as the publication of real-time open data.

Common southbound interfaces. For IoT device vendors and manufacturers it should become easier to offer suitable device stacks for integrating heterogeneous IoT components into a common environment, together with a market place for compliant IoT products and solutions.

Market place enablers. These should encourage sharing of urban IoT data and other relevant data sets among different stakeholders. By providing a market place as a one-stop-shop, it will become much easier for data consumers to discover and access urban data sources. The availability of a trusted market place with monetization mechanisms will allow third parties to generate easier revenue streams from their urban data sources. This will encourage more businesses to share currently closed data sources or incentivize deployments of new IoT infrastructure as secondary revenue streams can be generated, making more business cases viable. Data consumers may not require lengthy negotiations of license terms as data license terms can be negotiated from pre-configured options of the provider on the fly.

8.5.4 SynchroniCity Applications

Based on the shared architecture, SynchroniCity will deliver IoT-enabled urban services in the eight reference zone cities. The services will be developed and delivered in an ambitious two-stage approach.

First phase consists of three *initial applications*, based on the highest priorities among the members of the global Open & Agile Smart Cities initiative, a network of more than 100 cities worldwide which includes the SynchroniCity reference zones:

- Community Policy Suite
- Context-adaptive traffic management
- Multi-modal transportation

These applications are fairly standard services nowadays. However, there are still significant gaps when it comes to actually delivering them at scale, based on digital single market principles. Not all the SynchroniCity cities are involved in the deployment of all the initial applications.

The following phase is called "enrichment of the eco-system", and it adds another wave of *new applications* to the SynchroniCity portfolio, based on an open call where new companies and cities may propose applications and services built on top of the SynchroniCity architectural principles.

Figure 8.18 Interactive light art in Eindhoven.

A total of 3 million Euros has been allocated for the new applications, and the call for participation will launch in the spring of 2018.

Taken together, the applications developed and deployed at large scale in SynchroniCity will form a substantial contribution to a global market for IoT-enabled urban services.

8.5.5 Impact Creation

SynchroniCity consists of eight core reference zones in as many cities, but the ambition is to go well beyond this starting point: beyond the reference zone to the entire city, beyond a single silo to span multiple domains, and to go beyond Europe.

To reach this goal, the project was founded on a strong partnership basis with existing initiatives and communities, and has defined a set of ambitious KPIs.

At the core of SynchroniCity is the Open & Agile Smart Cities initiative (OASC)[4] which is a global network of national networks of more than

---

[4]www.oascities.org

100 cities in 23 countries that are working together to contribute to the establishment of a simple set of mechanisms for technical interoperability and comparability based on the needs of cities, i.e. the demand-side in the market. By adding global partners in Mexico, USA, Korea and Brazil, SynchroniCity spans a large diversity which is a characteristic of a global market. Potentially, the entire OASC network can easily adapt and deploy the SynchroniCity applications. Figure 8.15 shows the European reference zones, the global partners and the OASC cities.

A key focus of SynchroniCity is to contribute to the development of common specifications and ultimately standards. As shown in Figure 8.19 below, SynchroniCity takes input not only from OASC and the cities directly involved in the project, but also from other large initiatives, including prominently the FIWARE[5] initiative and the European Innovation Partnership on Smart Cities and Communities (EIP-SCC)[6]. Through targeted activities and deliverables as well as participation directly in the specification and standards development, SynchroniCity is contributing directly to a number of streams, including the ETSI Industry Specification Group on Context Information Management (ETSI ISG CIM)[7], the ITU-T Focus Group on Data Processing and Management for Smart Cities and Communities (FG-DPM)[8], and ISO/TC 268 on Sustainable cities and communities[9]. The project actively supports and contributes to the UN Sustainable Development Goals[10].

By having the open call where actors outside of the SynchroniCity project consortium are invited to enrich the eco-system, SynchroniCity actively seeks to facilitate support and impact beyond the closed group of initial partners.

8.5.6 Conclusions and Outlook

SynchroniCity has a clear ambition to deliver local value in cities and communities based on the global dynamics of digital connectivity, innovation power and capital. So far, efforts to create a global multi-sided market based on demand-side needs have not been successful. With the approach described above, the partners in the SynchroniCity consortium propose an approach

---

[5] www.fiware.org

[6] www.eu-smartcities.eu

[7] https://portal.etsi.org/tb.aspx?tbid=854&SubTB=854

[8] http://www.itu.int/en/ITU-T/focusgroups/dpm

[9] https://www.iso.org/committee/656906.html

[10] https://sustainabledevelopment.un.org/?menu=1300

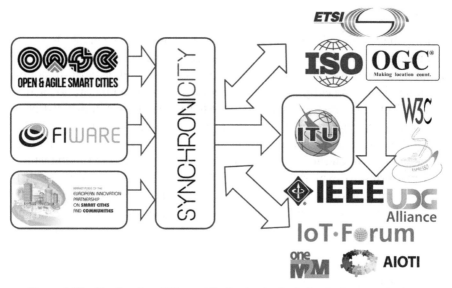

Figure 8.19 The SynchroniCity contribution to standards developing organisations.

which leverages interests from both sides of the market, with a clear focus on having humans in the centre. Together with the rest of the IoT Large Scale Pilot projects, SynchroniCity will hopefully bring Europe and the world a step closer to a well-functioning market, based on human needs and life between the systems.

8.6 AUTOPILOT – Automated Driving Progressed by Internet of Things

"AUTOmated driving Progressed by Internet Of Things" (AUTOPILOT) is a three-year project that started in January 2017, receiving funding from the European Union's Horizon 2020 research and innovation programme. The AUTOPILOT consortium, consisting of 45 partners, represents all relevant areas of the Internet of Things (IoT) eco-system. Its overall objective to enable safer highly automated driving through smart and connected objects – the IoT[11].

---

[11]Fischer, F. & Corazza, F. (2017).

8.6.1 Project Overview

During the last decade, numerous IoT technologies have been developed by the research community, including IoT software engineering tools and techniques, schemes for safeguarding security/privacy as well as infrastructures. Built upon these recently finished or ongoing research and innovation activities, AUTOPILOT focuses on utilising the IoT potential for automated driving and on making data from autonomous cars available to the IoT. In particular, AUTOPILOT aims to bring together relevant knowledge and technology from automotive and IoT value chains in order to develop IoT-architectures and platforms to explore the growing market for mobility services. IoT-enabled autonomous cars are tested, in real conditions, at six large-scale sites, whose test results will allow multi-criteria evaluations (technical, user, business, legal) of the IoT impact on pushing the level of autonomous driving.

8.6.1.1 Objective

Connectivity and the ability to collect data from thousands of objects surrounding vehicles are key enablers for highly automated driving. The IoT provides the mechanisms and tools to create virtual objects in the Cloud from real connected objects, thereby allowing these objects to become more automated in the not-so-distant future.

Overall, AUTOPILOT pursues five main objectives. First, it seeks to enhance the vehicle's understanding of its environment, utilising IoT sensors. Second, the project sets out to foster innovation in automotive IoT and mobility services. Third, it wants to use and evaluate advanced vehicle-to-everything connectivity technologies. Fourth, AUTOPILOT is designed to involve users, public services and businesses to assess the IoT's socio-economic benefits. Last, the project aims to contribute to the standardisation of IoT eco-systems worldwide.

More concretely, AUTOPILOT proceeds in four steps (see Figure 8.20). On large-scale test sites, data is collected in an IoT eco-system, defined as "objects in the physical world, which are capable of being identified and integrated into communication networks". Next, this data is collated and

Figure 8.20   Project flow.

analysed in an IoT Platform, defined as "interconnecting things based on existing and evolving interoperable ICTs", which is specifically dedicated to Automated Driving (see Section 8.6.2.1). The IoT Platform is then used for four use cases to test applications on Automated Driving (see Section 8.6.2.2). Ultimately, AUTOPILOT suggests some services, which can serve as business models for translating IoT Applications (see Section 8.6.2.3) into tangible offers for end users of automated driving.

8.6.1.2 Partners

The AUTOPILOT consortium brings together 45 partners from multiple countries and diverse backgrounds. The consortium is a balanced mix of organisations working in IoT as well as in Automation. This cooperation is an asset, because both industries have a very different approach to data sharing and vehicle control. IoT devices are by definition shared and open while automotive electronics are closed and safety-critical. The partner constellation in AUTOPILOT helps to overcome this problem and thus fosters implementation of IoT solutions in future vehicle architecture. The idea behind the consortium is to combine a wide range of knowledge areas to foster the project's effectiveness in attaining its objectives. Lastly, the AUTOPILOT consortium provides an avenue for authorities into the project. City authorities of the project sites are as much part of the consortium as are road operators.

Overall, the AUTOPILOT consortium can be characterised in four clusters: (1) those developing AD vehicles, (2) those developing IoT and networks, (3) those collecting data to evaluate the systems and their potential impacts, and, (4) those potentially developing innovative services based on the results. This comprehensive mix of partners is a key to the project.

8.6.2 Project Approach

AUTOPILOT follows a two-track approach, composed of both pragmatic and conceptual platform development. On the one hand, the pragmatic architecture serves not only as starting point for the development of the conceptual and unified platform but also as starting point for the use cases at the different Pilot Sites. On the other hand, the conceptual architecture is deployed very early and can be used by any new development. The conceptual architecture is to be used as the reference for discussion and collaboration with other Large Scale Pilots. To swiftly achieve tangible results, AUTOPILOT utilises pragmatically existing systems or other available installations.

In addition, AUTOPILOT adopts an iterative approach for innovation activities, known as Iterative and Incremental Development (IDD). The main rationale for IID in the design phase is to allow the project to better adapt to changing contexts, requirements, technological developments and boundary conditions. Therefore, the requirements are not frozen at the early stage of the project but will be reviewed based on preliminary evaluation results.

A maximum of three cycles is foreseen, whereas the need for applying all cycles is likely to be different for the partners and Pilot Sites, according to their level of maturity in IoT and automated driving at the start of the project. The multiple cycle approach offers all partners the necessary level of flexibility concerning their participation in the project as well as harmonising their contributions to the project.

The specific motivation for this approach in AUTOPILOT is threefold: First, there is a technical justification. One needs to merge the need to integrate many different components, with the need to start from existing solutions and adapt to users' early feedbacks. Second, there is an organisational justification, as Pilot Sites will learn from each other between cycles. Third, there is a business justification, as initial solutions to show on the demand side can pointedly visualise the demand side to IoT solutions for automated road transport.

8.6.2.1 AUTOPILOT's IOT Platform

AUTOPILOT's IoT Platform is held responsible on six parameters of design.

- Standard-Based: Legacy or proprietary IoT systems can be integrated into a common system;
- Abstraction: Abstraction-based information model following established High-Level IoT Architecture;
- Federation: Private and public IoT systems can co-exist and collaborate on-demand;
- Semantic: Based on formal semantics to achieve automated processing of large variety of information;
- Functional Distribution: Architecture enabling functional elements to be allocated into systems;
- Security: Provision of security functions and implementation of the "Privacy-by-Design" principles.

Today's IoT installations are mainly *Intranet-of-Things* meaning closed systems developed for a single commercial purpose and all following the same design principles. But the vision of AUTOPILOT is calling for an open IoT system, in which many participants are publishing and consuming IoT data and services. This is alike the Internet in which many participants can operate their own Web site. As shown with traditional Web-based systems, there is a need for aggregation and federation services, supporting developer and end-user with high-quality services (e.g. Google search, hotel booking engines, news aggregation services).

Those services are especially needed in the IoT for autonomous driving areas. Automated driving service operators will prefer to use a single service, delivering high quality and reliable information, rather than having to interface with thousands of information sources and service providers. This is called the IoT Federation, where autonomous systems agreed to collaborate by exchanging data and providing services to each other. The AUTOPILOT conceptual architecture provides such federation mechanism following the federation design principle. For example, for data access the FIWARE IoT Broker provides a large-scale federation infrastructure. Federation enables a finer grain control over IoT information compared to centralised IoT clouds. Federation also enables privacy-by-design thanks to clear policies and control points for information access.

Today's autonomous driving applications are relying on autonomous vehicle systems in which the needed information is gathered, processed and analysed on the vehicles themselves. As a consequence, the vehicle platform can base its driving and manoeuvring decisions solely on local information (Autonomous Car Zone). In the past year, newer approaches for cooperative driving have emerged, where driving functions take information gathered from other devices into account, e.g. using Car2Car or Car2Infra approaches (Cooperation Zone). With the emergence of the IoT, autonomous driving services can utilise IoT devices from the surrounding as well as services from the Cloud

8.6.2.2 Project Sites and Applications

AUTOPILOT employs large numbers of vehicles under normal traffic conditions, for which trials are being conducted at six large-scale Pilot Sites (Table 8.2).

- The Finnish Pilot Site, located in Tampere provides different outdoor conditions, such as slippery intersection in winter time, low visibility for environment perception sensors due to fog.
- The French Pilot Site is located in downtown Versailles, close to the castle. Its goal is to provide mobility services for tourists, based on a small fleet of automated vehicles and dedicated to a car-sharing.

- The Italian Pilot Site, is a testing infrastructure encompassing the Florence-Livorno highway together with road access to the settlement around the Livorno seaport.
- The Korean Pilot Site in Daejeon focuses on road situation information at an intersection, which is challenging for automated vehicles because of the large number of obstacles to be encountered there.
- The Dutch Pilot consists of three different test areas, including the Eindhoven University campus, the automotive campus parking, and a 6 km stretch of the A270 motorway.
- The Spanish Pilot Site is located in Vigo, covering the city's main street. It provides an operational test on cooperative systems between different types of vehicles and for underground parking.

AUTOPILOT deploys four different automated driving use cases: Automated Valet Parking (AVP) is a driverless Automated Driving use case including on-street car drop-off, driving to and from a parking spot, forwards and backwards manoeuvring as well as on-street passenger pick-up. The IoT allows this use case to be a Level 4 scenario, since data and control from different infrastructure sensors is essential. IoT functions include routing (parking availability through sensors), localisation of obstacles (parking cameras) and even control decision making at the IoT Edge.

Table 8.2   AUTOPILOT project sites and applications

Country	City/Region	Valet Parking	Highway Use	Platooning	Urban Driving
🇫🇮	Tampere	✓			✓
🇫🇷	Versailles	✓		✓	✓
🇮🇹	Livorno-Florence		✓		✓
🇰🇷	Daejeon				✓
🇳🇱	Eindhoven-Helmond	✓	✓	✓	✓
🇪🇸	Vigo	✓			✓

Highway Use is a use case focused on Automated Driving on motorways from entrance to exit, on all lanes, incl. overtaking. The driver must deliberately activate the system, but does not have to monitor the system constantly. There are no requests from the system to the driver to take over when the systems in normal operation area (i.e. on the motorway). Depending on the deployment of cooperative systems ad-hoc convoys could also be created if V2V communication is available.

The use case for Platooning is an Automated Driving scenario where fully automated driving or driverless vehicles will join and drive in a platoon, with a leading vehicle in front. The driving mode is very similar to the Highway Pilot, however driving in a platoon requires the vehicle to use advanced V2V communications. Two variants of platooning will be deployed and evaluated in AUTOPILOT, an urban one and a highway one.

The Urban Driving use case is based on the ERTRAC "Fully automated private vehicle" representing the SAE level 5, where "The fully automated vehicle should be able to handle all driving from point A to B, without any input from the passenger."

8.6.2.3 Services the intersection of IoT and automation

As stipulated in its acronym, AUTOPILOT seeks to foster the progress of automated driving. For this purpose, it makes use of the innovation of IoT developments in the field. Throughout the project, members of the consortium will work to develop eight specific services that draw on the use case applications of IoT (see Table 8.3). These services fulfil a double function. On the one hand, they make the idea of automated driving progressed through the IoT tangible for end users. On the other hand, they suggest viable business models for stakeholders that will carry on the findings of AUTOPILOT even after it ended.

8.6.3 Project Impact

To provide quantitative and qualitative evidence of the added value of IoT technology for automated driving, all large-scale Pilot tests are evaluated using the established FESTA methodology. The added value is formulated in hypotheses on objectives, ambitions and impact, and is measured in Key Performance Indicators (KPIs) or metrics from several perspectives. Data analyses and (technical) evaluation will be executed simultaneously with piloting to provide immediate feedback on test executing and input for the next iteration of development.

Table 8.3   IoT based automated driving services

City chauffeur services for tourists

Adaptation of car-sharing, dedicated for tourist to visit cities and other remarkable locations, being out of the loop to follow a multimedia presentation of the site in the vehicle.

Automated driving route optimisation

Monitoring of vehicles and environment to optimise automated driving routes to autonomous driving cars, by redistributing the traffic along alternative paths towards available parking spots.

Real time car sharing

Optimisation of vehicle allocation through collection of end user needs and analysis of IoT platform data to suggest car sharing (pick-up/drop-off) possibilities.

Driverless car rebalancing

Rebalancing of car sharing vehicles, with automated driving vehicles moving driverless from their last drop-off zone towards another pick-up location, either alone or in a platoon.

HD maps for automated driving vehicles

Provision of High-Definition maps data base for Automated Driving vehicles, built on the data collected by all connected vehicles.

6th Sense driving

Provision of risk metrics contributing to road assessment programme to create specific metrics and services for road rating dedicated to the autonomous driving capabilities of roads.

Dynamic eHorizon

Update of IoT based HD map data service through data from other vehicles' sensors and from other sources in real time, enabling to factor in dynamic changes to the route.

Electronic driving license

Generation and storage of secured identity profile to allow dedicated access to any smart city platform.

AUTOPILOT will focus on using standard-based IoT to ensure deploying interoperable and replicable IoT platforms and architectures. It will verify the interoperability and functionality during the development phase. Security and privacy by design will be a key impact of AUTOPILOT. Additionally, AUTOPILOT will deploy and evaluate several business services contributing or using automated driving. In this, the project will have to assess possibly competing elements of user acceptance, such as the perceived usability, ethical aspects and liability.

Overall, AUTOPILOT will pilot the use cases in a real public and private environment, with the consent and strong support of the City councils, some being in the consortium. Services like car sharing, city chauffeur and automated valet parking are expected to increase citizens' quality of life. The Pilot Sites, with their different automated driving use cases and business partners are an instrument to attract new entrepreneurs for creating further business opportunities. AUTOPILOT will promote the IoT as key enabler for automate driving and the related new business models linked with the IoT eco-system.

8.7 CREATE-IoT Cross Fertilisation through Alignment, Synchronisation and Exchanges for IoT

CREATE-IoT is the coordination and support action involving all IoT European Large-Scale Pilots innovation actions projects, articulating altogether the IoT European Large-Scale Pilots Programme that is built around eight activity groups. Through an active participation of these activity groups, IoT Large-Scale Pilot projects are able to contribute to the consolidation and coherence work that is implemented by the CREATE-IoT and U4IOT. This is done by supporting the clustering activities defined by the Programme and addressing issues of common interest such as interoperability approach, standards, security and privacy approaches, business validation and sustainability, methodologies, metrics, etc.

The ultimate goal of the IoT European Large-Scale Pilots Programme and the coordination/collaboration activities is to increase the impact of the activities and development in the IoT Large-Scale Pilots on citizens, public and private spheres, industry, businesses and public services. The activity groups are key enablers for the identification of key performance indicators to measure progress on citizen benefits, economic growth, jobs creation, environment protection, productivity gains, etc. The coordination mechanisms implemented through the activity groups will help to ensure a sound coherence and exchange between the various activities of the IoT Focus Area, and cross fertilisation of the various pilots for technological and validation issues of common interest across the various use cases.

The issues of horizontal nature and topics of common interest, for all IoT Large-Scale Pilot projects, such as privacy, security, user acceptance,

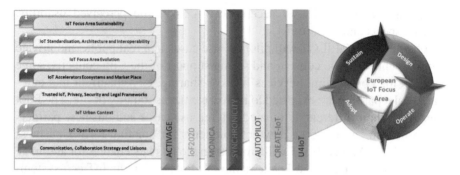

Figure 8.21 IoT European Large-Scale Pilots Programme Activity Groups.

standardisation, creativity, societal and ethical aspects, legal issues and international cooperation, etc., are coordinated by the activity groups (Figure 8.21) and consolidated across the pilots to maximise the output and to prepare the ground for the next stages of deployment including pre-commercial or joint public procurement.

The activity groups support and foster links between communities of IoT users and providers, with Member States' initiatives, and with other initiatives including contractual Public-Private-Partnerships (e.g. Big Data, Factories of the Future, 5G-infrastructure), Joint Technology Initiatives (e.g. ECSEL), European Innovation Partnerships (e.g. on Smart Cities), and other Focus Areas (e.g. on Autonomous transport).

The activity groups monitor that appropriate mechanisms are put in place so that pilots' impact can go beyond involved partners by also reaching external communities and stakeholders. Special attention should be given to those pilot projects which intend to launch open calls through cascade funding. Mobilising a wider community beyond the consortium is essential for the development of a secure and sustainable European IoT ecosystem. Beyond sustaining an ecosystem it is instrumental to include contributions to assure that the IoT infrastructures developed and implemented are viable beyond the duration of the Pilots.

The IoT European Large-Scale Pilots Programme coordination body led by the supporting and coordination actions, being CREATE-IoT part of it, will continuously monitor and adapt through the activity groups the common topics, challenges, best practices in order to maximise the expected impact of the IoT Large-Scale Pilots and coordination actions as outlined below:

- Validation of technological choices, sustainability and replicability, architectures, standards, interoperability properties and key characteristics such as security and privacy;
- Exploration and validation of new industry and business processes and innovative business models validated in the context of the pilots.
- User acceptance validation addressing privacy, security, vulnerability, liability and identification of user needs, concerns and expectations for the IoT solutions
- Significant and measurable contribution to standards or pre-normative activities in the pilots' areas of action via the implementation of open platforms
- Improvement of citizens' quality of life, in the public and private spheres, in terms of autonomy, convenience and comfort, participatory approaches, health, lifestyle and access to services.
- Creation of opportunities for entrepreneurs by promoting new market openings, providing access to valuable datasets and direct interactions with users, expanding local businesses to European scale.
- Development of secure and sustainable European IoT ecosystems and contribution to viable IoT infrastructures beyond the pilot lifetime.
- Ensure efficient and innovative IoT take-up in Europe, building on the various parts of the initiative (pilots, research, horizontal actions).
- Efficient information sharing across the programme stakeholders for horizontal issues of common interests.
- Extension and consolidation of the EU IoT community, including start-ups and SMEs.
- Validation of technologies' deployment and replicability towards operational deployment.
- Validation, in usage context of most promising standards and gap identification.
- Strengthening of the role of EU on the global IoT scene, in particular in terms of access to foreign markets.

8.7.1 Introduction

CREATE-IoT brings together 19 partners from 9 European countries. The project objectives are to stimulate collaboration between IoT initiatives, foster the take up of IoT in Europe and support the development and growth of IoT ecosystems based on open technologies and platforms. This requires synchronisation and alignment on strategic and operational terms through

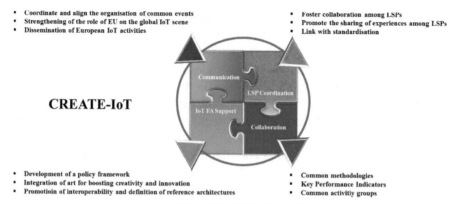

- Coordinate and align the organisation of common events
- Strengthening of the role of EU on the global IoT scene
- Dissemination of European IoT activities

- Foster collaboration among LSPs
- Promote the sharing of experiences among LSPs
- Link with standardisation

- Development of a policy framework
- Integration of art for boosting creativity and innovation
- Promotioin of interoperability and definition of reference architectures

- Common methodologies
- Key Performance Indicators
- Common activitiy groups

Figure 8.22   CREATE-IoT activities.

frequent, multi-directional exchanges between the various activities under the IoT Focus Areas as depicted in Figure 8.22. It addresses cross fertilisation of the various IoT Large Scale Pilots for technological and validation issues of common interest across the various application domains and use cases. The project fosters the exchange on requirements for legal accompanying measures, development of common methodologies and KPIs for design, testing and validation and for success and impact measurement, federation of pilot activities and transfer to other pilot areas, facilitating the access for IoT entrepreneurs/API developers/makers, SMEs, including combination of ICT and art.

CREATE-IoT takes a holistic approach in facilitating the exchange between the different IoT Large-Scale Pilot projects and providing the appropriate platform and tools that will enable fruitful cross-fertilization between IoT Large-Scale Pilots.

8.7.2 Conceptual Approach

CREATE-IoT aims to provide support to the different activities covered by the IoT Focus Area through the development of a coherent strategy for open exchanges and collaboration between the various actors involved. CREATE-IoT is focusing on promoting and discovering synergies across the different axes of the IoT Focus Area, as shown in Figure 8.23. The horizontal nature of (personal) data protection, security, user acceptance, standardisation, interoperability, creativity, societal and ethical aspects, legal issues and international cooperation are coordinated across the whole IoT ecosystem. Special attention is paid to the Large-Scale Pilots, but also to other initiatives

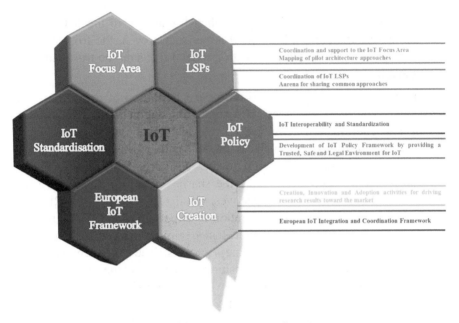

Figure 8.23 CREATE-IoT project axes.

in the IoT Focus Area, and other related initiatives at EU and global level involving different IoT communities, technology and innovation platforms, etc. The project boosts synergies, experience and knowledge sharing towards the creation of an inclusive IoT ecosystem, by defining a set of best practices and business models that will facilitate the replication of results and, thus, improve the overall impact of the parallel on-going activities, in particular the IoT Large-Scale Pilot projects. In summary, CREATE-IoT is working on:

- Ensuring coherent exchanges between the various activities of the Focus Area, and cross fertilisation of the various pilots for technological and validation issues of common interest across the various use cases.
- Supporting and assessing the current IoT Large-Scale Pilot projects, sharing best practices and aligning horizontal issues.
- Developing a more inclusive ecosystem where innovation, art and creativity take active part on it.
- Proposing a framework for the coherent integration of the EU IoT value chain, strengthening the links between different on-going initiatives in the IoT domain.
- Fostering interoperability of existing and future IoT solutions.

- Implementing and developing a policy framework in the IoT domain addressing the current horizontal issues that prevent the from massive deployment of IoT solutions, with a special focus on the trust and legal domains.

The project methodology is based on a clear definition of the current IoT ecosystem. The work provides the overall foundation for common inter-linking and coordination activities: all relevant stakeholders, technologies, reference architectures and business models will be the base over which the project will grow. The support to the IoT Large-Scale Pilot projects, as addressed by the project, fosters the sharing of experiences both from technical and non-technical perspectives. On the technical side, CREATE-IoT is focusing mainly on the federation of experiments and IoT reference architectures, whilst the non-technical side aims at sharing experiences, assessment methodologies and business models. The project exploits the synergies between arts and IoT and position creativity and innovation as a key driver and enabler for the progress of IoT. The methodology resulting from this work is tested in the different LSPs. The project addresses the creation of a framework for an integrated European IoT Value Chain that will grow the links between communities of IoT users and providers, as well as with Member States' initiatives and other related initiatives in the IoT domain.

The development of a Policy Framework will help addressing the multiple existing challenges, especially in the security and privacy domain, which are hindering the wide adoption of IoT solutions. The requirements for a trusted IoT label are under discussion. This label would better position EU IoT solutions in the global markets, being applicable not only to the whole IoT Focus Area, but also to other IoT domains. CREATE-IoT works on supporting industrial consensus building for implementing pre-normative and standardisation, coordinating the activities with Standard Developing Organizations (i.e. ISO, ETSI, CEN/CENELEC, W3C, IETF, ITU, IEEE, OGC, etc.) and other various IoT Global Alliances (AIOTI, IIC, Thread, AllJoyn, Open Connectivity Foundation (OCF), etc.), thus aligning activities at the national, European and global level.

8.7.3 Impact

CREATE-IoT provides the framework for supporting and coordinating the IoT Focus Area, in order to foster the take up of IoT in Europe and enable the emergence of IoT ecosystems supported by open technologies and platforms.

This is done through the coordination of complementary activities structured around IoT Large-Scale Pilots

The project ensures the framework environment and provides tools to ensure consistency and linkages between the pilots, complementing them by addressing ethics and privacy, trust and security, respect for the scarcity and vulnerability of human attention, validation and certification, standards and interoperability, user acceptability and control, liability and sustainability. The project implements a coordination body that ensures an efficient interplay of the various elements of the IoT Focus Area and liaises with relevant initiatives at EU, Member States and international levels.

For today's and tomorrow's IoT products, services, applications, solutions infrastructure and ecosystems, it is crucial that the society, its citizens and other (potential) customers and users have trust in what they use, buy, wish to enjoy or are connected to. In order to create a workable level of such trust and therewith durable adoption, one will need to have comfort, the offer will need to be credible and usable. In that scope, concepts such as ethics, safety, accountability, security and privacy by design have to be widely spread from the early stages of development for IoT products, services, applications, solutions infrastructure and ecosystems. To enable large scale deployments of IoT systems ensuring massive and durable user adoption, it will be essential that IoT and related IoT ecosystems are based on complementary architectures based on similar principles, enabling to leverage across multiple use cases and to catch the extra value arising from information exchange across multiple sectors/domains.

The goal of the CREATE-IoT project is to put in place an approach that can federate the IoT Large-Scale Pilots and, beyond them, other IoT and related ecosystems that are deployed worldwide. Given that IoT applications are meant to affect the living environment with minimum involvement from the human side, security breaches may have direct impact on human safety, and the privacy may be compromised by communications that happen without being noticed. Furthermore, security in IoT environment is more challenging to be achieved than in the traditional Internet of computers, due to power, bandwidth and computing constraints. Therefore, it tends to be neglected in early deployments despite its high criticality. In addition, the privacy risk requires specific overall horizontal attention across initiatives, because exposure of multiple anonymized data about the life, from home energy consumption to public transport infrastructure usage and office activities, facilitates link ability to individual identity, eventually compromising the privacy of each anonymous data. These aspects deserve specific attention in the supervision activities. There are many new ethical, legal and related questions and queries that need to be structured, addressed, clarified and sometimes rethought. There are many layers in any IoT vertical and horizontal matters such as for instance (personal) data protection, data management, security, product liability and net neutrality rise not once, but actually several times per layer. There are quite some layers necessary in order to build an IoT vertical, let alone the many cross-vertical data value chains and liabilities that need to be addressed.

CREATE-IoT has an important and critical impact on potential barriers, on providing the stakeholders with multi-angle and sufficient knowledge, with a trusted environment for the development of IoT, and with comprehensive technical and non-technical practical landscaped guidelines, mechanisms, best practices and other legal and related knowledge, and on giving support to projects across the IoT Focus Area. The ultimate aim is to create the Trusted IoT concept.

CREATE-IoT's expected impact is in ensuring that start-ups, developers and SMEs are able to play a role in the new IoT paradigm. One of the requirements is that the platforms deployed across the IoT Focus Areas must be open to enable sustainability and continued innovation and development of new cases. By ensuring that there are appropriate support mechanisms for SMEs and start-up participation, CREATE-IoT is working to:

- Enable start-ups and SMEs to engage with the IoT Large-Scale Pilot projects and build the skills and technology base they need to be part of the IoT ecosystem.
- Provide clarity about legal considerations on privacy and other related topics.
- Invite disruptive innovation from stakeholders who have no proprietary technology or legacy considerations.
- Ensure that the voice of smaller and new businesses across Europe can be heard in the debates on standards and interoperability.
- Lay the foundations for truly open (although not necessarily free of charge) platforms that withstand changes in any given ecosystem.

The longer-term impact of the support to SMEs is seen in Europe's ability to compete on the world stage with new companies that can contribute to the creation of skilled jobs and economic growth. CREATE-IoT creates impact through the development and implementation of a methodology to integrate artistic practices and creation in IoT innovation, adoption and market penetration. The exploitation of the combination of ICT and the arts has direct impact on IoT Large-Scale Pilots and Focus Area by enhancing a bottom-up user adoption and introducing a critical approach to technology in the design, development, dissemination of new products, technologies, services and experiences in various IoT application domains.

The expected impact resulting from the combination of ICT and the arts can be summarized in few points:

- Increased user adoption and confidence in IoT systems and IoT Large-Scale Pilots thanks to the consideration of user-centric methodologies and consumer-citizens assessment while emerged in IoT systems and experiences.
- Boost of socially driven innovation processes in IoT thanks to co-creation methodologies and a solid critical approach enabled by artists attributing meanings to technologies.
- Enhanced promptness to innovate and enhanced EU ICT competitiveness driven by the emerging hybrid field of ICT and the Arts supported by the EU Digital Agenda STARTS initiative.
- Consolidation of an EU IoT Art-Science cluster of practitioners that can extend the impact of CREATE-IoT beyond its time frame and scope.

8.8 U4IoT – User Engagement for Large Scale Pilots in the Internet of Things

This chapter presents the U4IoT Coordination and Support Action which aims at supporting end-user engagement in the five European Large-Scale Pilots. It comprises a short introduction of the project objectives followed by a review of the various strands of work aimed at achieving those objectives including enshrining participants' right to privacy and the protection of their personal data.

8.8.1 Introduction

U4IoT will support the Large Scale Pilots (LSPs) funded by the European Commission in the context of the Horizon 2020 research programme. It will enable a citizen-driven process by combining multidisciplinary expertise and complementary mechanisms from state-of-the-art European organisations. It will also analyse societal, ethical, and ecological issues related to the pilots in order to develop recommendations for tackling IoT adoption barriers including educational needs and skill-building.

U4IoT combines a wide range of knowledge and experience from a number of leading European partners in end-user engagement through crowd-sourcing, Living Labs, co-creative workshops, and meetups designed to support end-user engagement in the Large Scale Pilots. Its strategy is built on four main sets of activities. It will:

- Develop a toolkit to facilitate the LSPs' end-user engagement and adoption activities including: online resources and tools for end-user engagement; privacy-compliant crowdsourcing & crowdsensing tools and surveys to assess end-users' acceptance of the pilots; online resources and an innovative game that will promote awareness of privacy and personal data protection risks with guidelines on personal data protection;
- Support and mobilise: end-user engagement through the training and supporting of the LSP teams so that they may organise their own co-creative workshops; meetups; training on the use of crowdsourcing & crowdsensing tools in an efficient and privacy-friendly way in line with IoT Lab[12] tools; the presentation and facilitation of Living Labs support; an online pool of experts for end-user engagement; online training modules.
- Analyse societal, ethical, and ecological issues related to the pilots' end-users and make recommendations based on the analysis of IoT adoption barriers and how to tackle them including education and skill-building. It will leverage on end-user interactions to design participatory sustainability models that can be replicated across LSPs and future IoT pilots.
- Support communication, knowledge-sharing, and dissemination including: the development of an interactive website with an online toolkit as well as online knowledge database on lessons learned, FAQs, solutions, and end-user feedback. It will support the end-user communication and outreach strategies for LSPs and will enable information-sharing and retro-feed towards LSPs and their end-users.

8.8.2 Engaging End-Users throughout the Life of the LSPs

One of the key objectives of U4IoT is to support and promote end-user engagement across the entire lifecycle of the Large Scale Pilots from design, to implementation, through to assessment. More specifically, the project will encourage and support:

---

[12]www.iotlab.eu

End-user engagement in the design phase of the LSPs

U4IoT will provide methodologies, such as the co-creative workshops, that will enable pilots to involve the local end-users from the beginning of the project.

End-user engagement in the implementation phase of the LSPs

U4IoT will provide a complete set of tools to actively engage end-users in the design and implementation of the pilots' products and services.

End-user engagement in the exploitation and assessment phase of the LSPs

Privacy-friendly crowdsourcing and survey tools will enable the pilots to monitor end-user perception and acceptance of their projects during the implementation phase. It will enable the pilots to identify issues that need to be addressed so that they may maximise end-user acceptance and satisfaction.

This comprehensive approach is illustrated in Figure 8.24 below.

8.8.3 Embedding Personal Data Protection by Design

U4IoT will adopt a privacy-by-design approach. It will comply with the new General Data Protection Regulation as well as other complementary

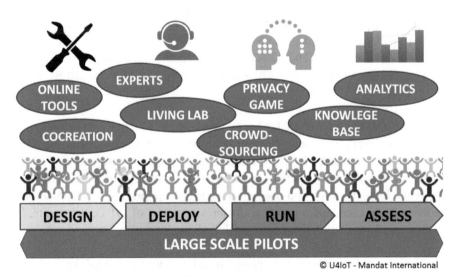

© U4IoT - Mandat International

Figure 8.24 U4IoT overall concept.

international (UN-, ITU-, and OECD-related norms) and European standards. These include the Charter of Fundamental Rights of the EU (arts. 7 and 8), the Treaty of Lisbon (arts. 16, 39, 88, and 169), Directives 95/46/EC, 97/66/EC, 2002/58/EC, 2002/21/EC, 2009/140/EC, 2009/136/EC, and Regulation(EC) N°45/2001.

U4IoT not only features expertise in personal data protection, but also has experts on gaming methodologies. One intended outcome will be the creation of a game whose aim is to raise awareness on privacy in connection with the Internet of Things. This will help to better connect with end-users and raise awareness of personal data protection issues and challenges they will face as the technology achieves greater social penetration.

8.8.4 Developing an Ad Hoc Toolkit for End-User Engagement

U4IoT is developing a customised toolkit for end-user engagement in the Large Scale Pilots, including:

Online resources for pilots including tools for end-user uptake
U4IoT will identify, gather, and provide relevant tools and online resources for end-user engagement including strategic guidelines and steps for user engagement and a pool of methodologies and tools to be used in different states or steps in product and service co-design and co-creation. U4IoT will provide resources supporting LSPs in engaging end-users in their pilots. The online resources will be hosted on the project website. In addition, U4IoT will compile these resources and advice into a practical handbook for the Internet of Things pilot deployments. The handbook will also be designed to serve future Internet of Things pilots.

Crowdsourcing and survey tools
U4IoT is leveraging and customising the crowdsourcing and crowd-sensing tool developed by the European research project IoT Lab (http://www.iotlab.eu). It enables researchers to collect fully anonymised input and feedback from end-users through a smart phone application in order to assess, fine-tune, and validate end-user acceptance of IoT deployments. It is particularly well-suited for large-scale IoT deployments.

Guidelines and an interactive game for privacy and personal data protection
U4IoT will develop a game to educate and raise awareness on IoT and privacy issues among the LSPs' stakeholders and their end-users. Our goal

is to develop a game that fully integrates some basic concepts of privacy (mainly online and IoT privacy) into a working and enjoyable mechanism through which understanding such concepts becomes crucial to succeeding in the game. Different versions of the game will be considered and an interactive game will be developed in close collaboration with the partners of the project and end-users. The game will highlight key concepts of personal data protection such as: a priori informed consent, profiling, data processing, anonymity, pseudonymity, etc. The game will focus on privacy issues related to Smart Cities and the Internet of Things.

U4IoT website and dissemination
All presentations, documents, online tools, knowledge database, etc. will be made available and maintained on the U4IoT website. Not only will it document the activities of the project, but will also provide tools for interaction with relevant projects and users. The website will be used as a source for other promotion and dissemination channels such as social networks, which will complement end-user engagement activities done "out and about". To this end, the "mood of the city" kiosk will be used as a means of attracting interest and spreading information about the project and related activities.

Online knowledge database on lessons learned, solutions, and user feedback
Another important element of the project is the development of a knowledge base on lessons learned, solutions, and user feedback. It will be maintained and regularly updated in order to capitalise on experiences and lessons learned. At the end of the project, the knowledge base and online platform will be formally transferred to the IoT Forum who has agreed to maintain it.

8.8.5 Supporting and Mobilising End-User Engagement

Beyond the development of tools, U4IoT considers that direct support to the Large Scale Pilots is required as a complementary measure to enable effective end-user engagement. A set of support actions and services are being developed:

Co-creative workshops
Co-creative workshops can support end-user engagement in the early stages of the LSPs by co-designing solutions in a multi-stakeholder setting. During the workshop, collective knowledge of end-users, local stakeholders, and

experts from LSPs is exchanged, combined, and captured. The co-creative workshop methodology and toolkit enables experts to empathise with the needs of end-users while end-users are empowered to communicate on an expert level. The workshop program results in one or more use cases that are co-created by the participants.

Analysing and converting the qualitative data gathered throughout the workshop leads to a variety of functional requirements that are based on end-user needs, however, designing, organising, facilitating, analysing, and implementing co-creative workshops in LSPs can be challenging. U4IoT therefore aims to offer a co-creative training program to increase end-user engagement during the beginning of many LSPs in the IoT-1 call.

In this training program, experts from different LSPs are invited to learn how to implement and utilise co-creative workshops in their projects. Experts will interact directly with end-users through the co-creative workshops, ideally increasing empathy and leading to more meaningful IoT-related solutions that are based on the needs of end-users. More LSPs adopting the co-creative workshop methodology will lower the threshold for involving end-users in future LSPs and improve our knowledge on applying this methodology in different contexts.

Living Labs methodology support

U4IoT will also provide the LSPs with Living Lab methodology support. Here, the LSPs will get access to guidelines and services online regarding how to apply Living Lab methodologies within the LSPs. Through this, the LSPs get guidelines on how to develop open and user-driven innovation ecosystems that support end-user engagement in the IoT pilots from the early phases and throughout the whole pilot. In relation to this, U4IoT will provide dedicated on-demand advisory support for the LSPs in Living Lab approaches. This methodology support will mainly be provided as an online tool with webinars and online supporting services.

Online pool of experts for end-user engagement

The U4IoT online training programme will assist LSPs to set up and deploy an effective end-user engagement strategy. The components of the programme consist of an interactive flow-diagram, e-courses, example materials, and an expert pool. The online pool will be a list of experts with details of their specific expertise. The pool will be available online and is part of a front-desk that also contains a list of end-user communities for open calls and a list of FAQs. The LSPs can use the list to contact experts directly for questions and answers or for custom advise and coaching on end-user engagement.

8.8.6 Recommendations on IoT Adoption and the Sustainability of IoT Pilots

Tackling IoT adoption barriers
U4IoT will start by analysing societal, ethical, and ecological issues related to the pilots with end-users and develop a set of corresponding recommendations. This will be achieved by analysing IoT adoption barriers and provide guidelines for tackling these barriers, including educational needs and skill-building.

Sustainability models for IoT pilots
U4IoT will design sustainability models for the LSPs. These models are participatory and leverage on end-user interactions by involving not only the LSPs themselves, but also their various stakeholders and target audiences, from companies to public organisations to citizens. The U4IoT sustainability models will be designed to help the LSPs and their stakeholders, but can also be replicated in future IoT pilot deployment throughout Europe."

8.8.7 Collaboration, Outreach, and Dissemination

A key factor of success for the project resides in its capacity to outreach towards the relevant stakeholder. The collaboration, outreach and dissemination strategy is threefold:

Cooperation strategy with the LSPs
U4IoT will work in close interaction with the LSPs and will offer its support to each LSP consortium from the beginning of the project. U4IoT will organise joint meetings and ad hoc training activities for all LSPs. In parallel, U4IoT will collaborate closely with the other CSA and the European Commission.

Cooperation strategy with the IoT Ecosystem and the IoT Forum
U4IoT will work closely with the IoT ecosystem and will contribute to the visibility and outreach of the LSPs. U4IoT will also work closely with the IoT Forum (chaired by MI), the AIOTI, the IERC, the ITU, and IEEE (more details in Section 8.2).

The IoT Forum will play an important role in supporting the interaction with the IoT research community, as well as in ensuring a long-term exploitation and maintenance of the U4IoT online tools and platform.

End-users outreach, engagement, and active involvement
U4IoT's main users are the consortia leading the Large Scale Pilots implementation. The prime focus of U4IoT is to enable active participation and engagement of final end-users in the pilots. Beyond the end-user engagement methodologies provided and developed in coordination with the pilots, U4IoT will also provide transparent information on its website with resources made available to final end-users too.

8.8.8 A Systemic and Cybernetic Approach for End-User Engagement

Through its aforementioned activities, U4IoT is following a fully integrated and systematic approach. It intends to follow and enable a virtuous cycle by collecting and internalising the feedback from the various stakeholders. The model adopted by U4IoT is summarised in Figure 8.25.

The analytical part is central and enables us to develop the knowledge, adapt and improve the support, and fine-tune the tools. These elements are then applied to the LSPs. The results are then monitored and analysed, closing the loop.

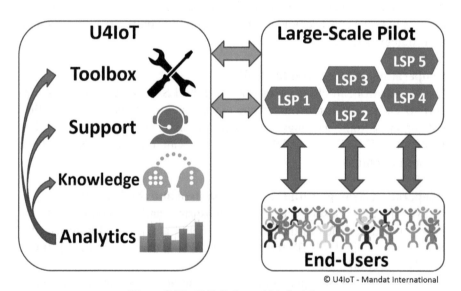

© U4IoT - Mandat International

Figure 8.25   U4IoT General Methodology.

8.8.9 Discussion

U4IoT stands as the primary support that the LSPs of the IoT-1 call will use to engage with their audiences and develop deep connections with their end-users. This will be achieved largely due to the extensive experience and knowledge base of the U4IoT partners, which they will provide to the LSPs both offline and online. Continuous support will be provided through our online portal and via an expert pool that will cover all the relevant topics and issues that the LSPs are likely to face in the realm of end-user engagement.

Expert knowledge will be accompanied and underpinned by technical tools including state-of-the-art crowdsourcing technology, by a robust legal framework, and by innovative engagement methodologies such as the privacy game and co-creative workshops. All of this will be achieved while simultaneously accounting for societal, ethical, and ecological considerations.

References

[1] European Commission: The 2015 ageing report, underlying assumptions and projection, methodologies (2014).

[2] United Nations: World population ageing 2013, Economic and Social Affairs (2013).

[3] Rechel B, et al. Ageing in the European Union (2013).

[4] The 2015 Ageing Report. European Commission, (2015). http://europa.eu/epc/pdf/ageing_report_2015_en.pdf

[5] Christensen K., et al. Ageing populations: the challenges ahead (2009).

[6] Smart Living Environment for Ageing Well Report. AIOTI WG5, October 2015.

[7] United Nations: World population ageing 2013, Economic and Social Affairs (2013).

[8] Rockwood K., et al. "The Canadian Study of health & ageing Clinical Frailty Scale", Can Med Assoc J 005; 173(5): 489–95. Dalhousie University, Canada

[9] H. Sundmaeker, C. Verdouw, SjaakWolfert, L.P. Freire, Internet of Food and Farm 2020, in: O. Vermesan, Friess, Peter (Ed.), Digitising the Industry River Publishers, 2016, pp. 129–150.

[10] C. Verdouw, Wolfert, S., Tekinerdogan, B., Internet of Things in agriculture, CAB Reviews, 11 (35) (2016) 1–11.

[11] C.N. Verdouw, J. Wolfert, A.J.M. Beulens, A. Rialland, Virtualization of food supply chains with the Internet of Things, Journal of Food Engineering, in press (2016).

[12] L. Pérez-Freire, L. Brillouet, M. Álvarez-Díaz, D. García-Coego, A. Jiménez, L. Murard, Smart Farming and Food Safety Internet of Things Applications – Challenges for Large Scale Implementations, in, AIOTI WG06, 2015, pp. 49.

[13] EIP-AGRI, Focus Group Precision Farming: Final Report, in, EIP-Agri/EC, 2015.

[14] J.W. Kruize, J. Wolfert, H. Scholten, C.N. Verdouw, A. Kassahun, A.J.M. Beulens, A reference architecture for Farm Software Ecosystems, Computers and Electronics in Agriculture, 125 (2016) 12–28.

[15] O. Vermesan and P. Friess (Eds.). Digitising the Industry Internet of Things Connecting the Physical, Digital and Virtual Worlds, ISBN: 978-87-93379-81-7, River Publishers, Gistrup, 2015.

[16] O. Vermesan, R. Bahr, A. Gluhak, F. Boesenberg, A. Hoeer and M. Osella, "IoT Business Models Framework", 2016, online at http://www. internet-of-things-research.eu/pdf/D02_01_WP02_H2020_UNIFY-IoT_ Final.pdf

[17] F. Fischer and F. Corazza, (2017), "Data on the Road", Baltic Transport Journal, No. 2, pp. 64–65.

9

A Smart Tags Driven Service Platform for Enabling Ecosystems of Connected Objects

Kaisa Vehmas[1], Stylianos Georgoulas[2], Srdjan Krco[3], Liisa Hakola[1], Iker Larizgoitia Abad[4], Nenad Gligoric[3] and Ingmar Polenz[5]

[1]VTT Technical Research Centre of Finland Ltd, Finland
[2]University of Surrey, UK
[3]DunavNET, Serbia
[4]EVRYTHNG LTD, UK
[5]Durst Phototechnik Digital Technology GmbH, Austria

Abstract

Internet of Things (IoT) is about connecting objects, things and devices and combining them with a set of novel services. IoT market is unstoppably progressing, introducing a lot of changes across industries, both from the technological and business perspectives. Optimization of the whole value chain is providing many opportunities for improvements leveraging IoT technologies, in particular if information about the products is available and shareable.

TagItSmart project is creating an open, interoperable set of components that can be integrated into any cloud-based platform to address the challenges related to the lifecycle management of new innovative services. TagItSmart is a three years project (2016–2018), consisting of 15 consortium partners from Europe. The project is funded under the Horizon 2020 program.

A main target of the TagItSmart are everyday mass-market objects not normally considered as part of an IoT ecosystem. These new smarter objects will dynamically change their status in response to a variety of factors and be seamlessly tracked during their lifecycle. This will change the way users-to-things interactions are viewed. Combining the power of functional inks with the pervasiveness of digital and electronic markers, a huge number of objects will be equipped with cheap sensing capabilities thus being able to capture new contextual information. Beside this, the ubiquitous presence of smartphones with their cameras and NFC readers will create the perfect bridge between everyday users and their objects. This will create a completely new flow of crowdsourced information that can be exploited by new services.

9.1 Introduction

Internet of Things (IoT) is about connecting objects, things and devices and combining them with a set of novel services. IoT market is unstoppably progressing, introducing a lot of changes across industries, both from the technological and business perspectives. The vision of the IoT did not change many from the beginning, but the reach that current technology brings to the table is still limited due to the certain limitations (i.e. technological limitation or for the practical reasons such as prices of the tags in some mass market scenarios) in different application of IoT.

Optimization of the whole value chain is providing many opportunities for improvements leveraging IoT technologies, in particular if information about the products is available and shareable. Many related industries are going to be affected by this, as packaging and insurance companies, among others, are requested to be more transparent to consumers, while consumers (78%) prefer brands that create unique and personalised content and are more interested in building a relationship with these companies [1].

Consumer packed goods (CPG) companies can prepare themselves for a range of possible futures by harnessing technology, reinventing brands, and exploring new business models [2]. The following five potential "undercurrents" that may impact the consumer product industry in 2020 are identified: 1) unfulfilled economic recovery for core consumer segments, 2) health, wellness and responsibility as the new basis of brand loyalty, 3) pervasive digitalization of the path to purchase, 4) proliferation of customization and personalization, and 5) continued resource shortages and commodity price volatility.

An important aspect to take into account is the need for a service economy around IoT [3]. The interconnection of products will promote ecosystems of new online services; therefore, new business models based on this network capital are gaining momentum. The new value chains will increasingly organize themselves as networks around consumers, offering a multiplicity of channels and interfaces across all value-add processes and business entities [4]. This makes consumer the one in charge, whose decisions affect the whole value-chain. Sharing information throughout the whole lifecycle of products and reactivity to context information are key in the short term.

TagItSmart project sets out to address the trends highlighted above by redefining the way we think of everyday mass-market objects not normally considered as part of an IoT ecosystem (Figure 9.1). These new smarter objects will dynamically change their status in response to a variety of factors and will be seamlessly tracked during their lifecycle. This will change the way users-to-things interactions are viewed. Combining the power of functional inks with the pervasiveness of digital (e.g. QR-codes, quick response codes) and electronic (e.g. NFC tags, near field communication) markers, a huge number of objects will be equipped with cheap sensing capabilities thus being able to capture new contextual information. Beside this, the ubiquitous presence of smartphones with their cameras and NFC readers will create the perfect bridge between everyday users and their objects. This will create a completely new flow of crowdsourced information, which extracted from the objects and enriched with user data, can be exploited by new services.

TagItSmart is creating an open, interoperable set of components that can be integrated into any cloud-based platform to address the challenges related to the lifecycle management of new innovative services capitalizing on objects "sensorization". To validate designed components and boost their adoption, a set of industrial use cases are used as a baseline for development, while additional stakeholders are being engaged through a co-creation Open Call approach.

TagItSmart is a three years project (2016–2018), funded under the part of Horizon 2020 program. The consortium consists of 15 partners coming from Austria, Finland, France, Italy, Netherlands, Romania, Serbia, Spain, Sweden, and United Kingdom. The coordinator of the project is DunavNET.

TagItSmart project aims to create a set of tools and enabling technologies with open interfaces that can be integrated into a platform of choice, enabling users across the value chain to fully exploit the power of mass-market connected products capable of sensing their own environment.

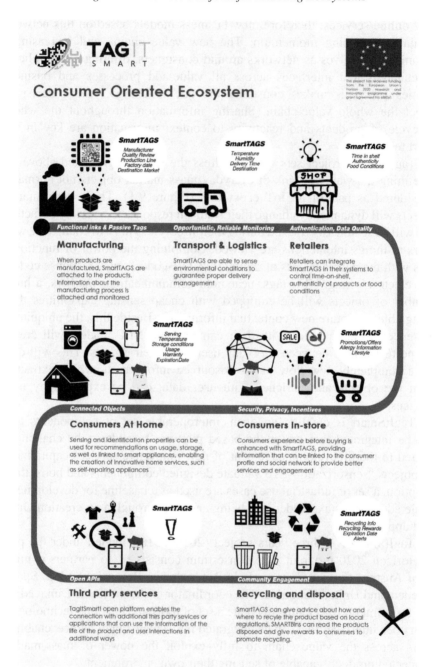

Figure 9.1   Basic concept of TagItSmart project.

9.2 Use Cases

The use cases for the TagItSmart project were created by combining expertise and real-life interests of consortium partners. In total five use cases were selected: 1) Digital product, 2) Lifecycle management, 3) Brand protection, 4) Dynamic pricing, and 5) Home services.

The digital product use case is a kind of umbrella use case including the whole value chain from manufacturer to transport, retail, consumer and recycling, see Figure 9.2. The other use cases concentrate more in details to one or two stages of the lifecycle of the product.

9.2.1 Digital Product, Digital Beer

Digital product use case extends a base use case for fast-moving consumer goods (FMCG) that want to become "smart" via SmartTag and TagItSmart. It combines novel solutions and enabling technologies and tools to create smart solutions for the whole value chain: manufacturer, transportation, retail, consumer and recycling. To implement the use case, a real FMCG has

Figure 9.2   The TagItSmart use cases under the Digital product umbrella.

Figure 9.3 The idea of Digital product use case.

been chosen. In this use case the beer and beer bottles are tracked from the manufacturer until recycling the bottle at the end of the value chain. From TagItSmart's digital product, digital beer needs at least functionalities for item-level control, life-cycle management, digital engagement and authenticity tracking. In addition, digital beer needs some basic sensing capabilities for monitoring conditions of the beer, see Figure 9.3.

The Digital product use case involves every part of the FMCG value chain and provides base functionalities for the TagItSmart project and other use cases. Making this use case live could affect every stakeholder in the chain and provide them new possibilities now and in the future.

By implementing this use case the manufacturers are able to control products that leave their factories throughout their lifecycle, e.g. where and how the products are transported and in which conditions, when they have been delivered to the retail stores and when sold to the consumers. This use case also creates a new channel for manufacturers to communicate with the consumer, so that they can enable easy access to related information on product and item-level which is not only static but depend on lifecycle and historical data of the item.

Especially consumer awareness has great potential when it comes to creating new business throughout the value chain; it makes it possible for product manufacturers to create products that address market needs better. Consumer awareness links to both two-way communication and item information, with two-way communication consumer can initiate conversion directly between him/her and the product manufacturer. This makes it possible to order customized products and removes overhead (and possible loss of information) when giving feedback of products consumed. However, it also works the other way; product manufacturer can reach customers with updated information about product and conditions it was manufactured, sources they use, etc.

In this use case, retail stores are not anymore midpoint warehouses but are more deeply engaged into the chain, giving them the opportunity to interact both with customers and product manufactures actively taking part in determining the next steps in product lifecycle.

Digital product use case also opens up possibilities for new use cases and new business. These solutions can be easily extracted to other fast-moving consumer goods. It also provides base functionalities for advanced use cases like home appliances, new recycling processes and customer-manufacturer engagement. These (and any other) use cases can use functionalities implemented by digital product use case as they are, extend them or replace them in order to support use case specific requirements.

9.2.2 Lifecycle Management

In Lifecycle management use case the aim is to implement a system/technology that allows the lifecycle management of every fast-moving consumer good (FMCG), or consumer packaged goods (CPG), that motivates and helps companies and citizens recycle their waste items, overcoming and solving current limitations and problems. The idea of this use case is described in Figure 9.4.

A key sustainability metric is the reduction in waste sent to landfill, but also the promotion of recycling. Items that are recyclable are often wrongly placed in a landfill bin, due to lack of awareness or engagement by the end user. This problem is compounded by the number of different regulations governing recycling at the local level and an inability to engage directly

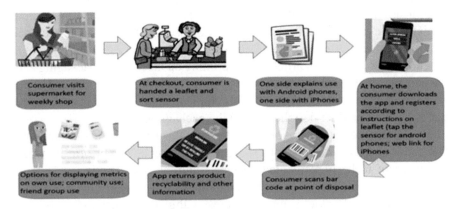

Figure 9.4 The idea of Lifecycle management use case.

with end users at the point of product disposal. An opportunity exists for a system that can give end users geographically relevant information on the recyclability of an item at this point in time, either within the home, or at a range of other recycling outlets.

9.2.3 Brand Protection

According to the report on EU customs enforcement of intellectual property rights (IPR) of the European Commission from 2015 for year 2014 [5], which is: a) Report on EU customs enforcement of intellectual property rights – Results at the EU border 2013; Publications Office of the European Union, 2014, Luxembourg and b) Report on EU customs enforcement of intellectual property rights – Results at the EU border 2014; Publications Office of the European Union, 2015, Luxembourg, 35.5 million enforcements of IPR by case were detected at the customs of the borders of the European Union with a total domestic retail value of 617.1 million Euro. Roughly, 47 % of the cases were reported to be from textile products such as clothing, clothing accessories and shoes, which is even ahead of medicines and cigarettes.

Counterfeiting, specifically in textile products but also expandable to other goods, is caused by several factors including the simplicity of the production and refinement processes, the cheap way of production and the low structure complexity which widely opens the doors for illegal copying. Brand protection strategies based on the implementation of barcodes, tags or labels to the product till nowadays had only a moderate success rate, whereas sophisticated anti-counterfeit systems (e.g. active radio-frequency identification (RFID) tags, electronically secured packaging) are more limited to high-value products. In the brand protection use case the aim is to develop a security platform for goods which includes both the direct digital printing of a functional QR code on an article or the printing on labels or on packaging and the generation of suitable coding and decoding environment that are deployed at the production process that is simply implementable by the manufacturer directly on the product periphery or on the packaging.

More specifically the aim of this use case is the development of functional QR codes and their readout for the purpose of a security platform for originality proof to be implemented on labels and tags. The basic mechanism of the security platform is a light- or temperature-induced colorization or decolourization of inkjet-printed textures within the functional code, see Figure 9.5. The encoding is performed via the data matrix approach (see use case digital product).

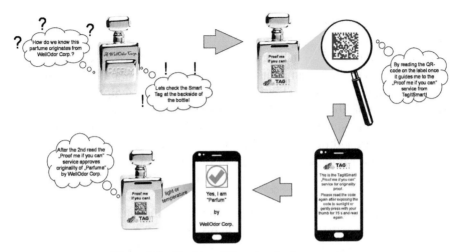

Figure 9.5   The idea of Brand protection use case.

9.2.4 Dynamic Pricing

The idea of the dynamic pricing use case is to deliver safe and fresh food, using packed meat as an example product that can benefit from conditions dependent pricing. Each package of meat is tagged with a SmartTag on the production line. The tag is capable of detecting temperatures higher than allowed for storing fresh meat. Once the temperature goes above the threshold for a certain period of time, the tag reacts and changes its content to indicate this event.

The SmartTags benefit several stakeholders in the value chain. Once the package arrives to a shop it can be scanned by the retailer. In that case, the packages that were stored in wrong temperature conditions will be removed. However, individual scanning of each package is not possible and the packages might end up on shelves. Once on the shelves, the consumers will be able to scan individual packages before buying. When consumers scan a package using a smartphone application they will obtain information about the meat: the origin, time of packaging, discount information and transport conditions. The obtained information is also forwarded to the meat supplier and potentially to other actors in the supply chain. At this point, it is possible to invoke condition dependent pricing mechanism depending on the relevant parameters like temperature in which the item was stored or time spent on the shelves.

Consumer takes the meat package to the cash register where it is scanned again. At this stage, the information is sent to the meat supplier and recorded in retailer's information system. In case the consumer had not scanned the item previously, the condition dependent pricing procedure is triggered in case the conditions for such procedure are met. After the checkout, the consumer goes to the car and puts the groceries in the trunk. When scanning is done at home, the SmartTag might indicate wrong temperature and the TagItSmart application would then suggest appropriate measures taking into account the time since the item was scanned at the checkout, i.e. the last control point. In addition to this, the consumer will also receive information about the recipes suitable for the type of purchased meat as well as instructions on how to dispose of packaging. The idea of dynamic pricing use case is presented in Figure 9.6.

Cold supply chains have been in use in various industries for a long time. The main characteristics of these systems is that the monitoring is done on a box or a pallet level while the items are in a truck, refrigerator etc. Modern supply chains solutions do not cater for monitoring at individual item level. Also, monitoring at critical points, when items are being handed over between different modes of transport or between two cold points (refrigerators) is lacking, thus preventing the stakeholders from having to have the complete picture and potentially exposing them to problems. Particularly critical is loading items to a transport vehicle and unloading at warehouse stores.

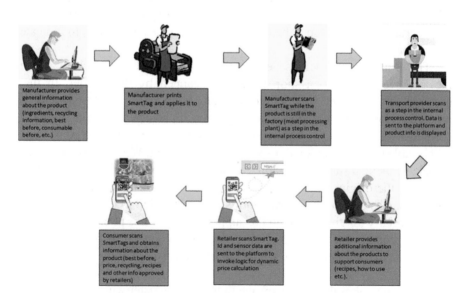

Figure 9.6 The idea of Dynamic pricing use case.

Integrated dynamic systems for monitoring products, the conditions of transport and storage, the relationship of its expiry date and price are not currently available. If a consumer is not a member of a loyalty programme, neither supplier nor retailer have the ability to interact with consumers in an easy manner and learn about the consumer habits. The goal of this use case is to enable monitoring of items at individual level, including at the loading/off-loading times. Further to this, the goal is to extend the monitoring of the item to its shelf time and further to consumer's house and eventually waste. Using this approach all stakeholders will be able to obtain information that can be used to improve work processes and increase quality of service.

9.2.5 Home Services

A good usage of the heating equipment (e.g. a boiler and an associated filtering station) is one means to provide comfort and a healthy living at home. This implies proper 1) maintenance, 2) repair, 3) replenishment, and 4) monitoring. This use case will test different services to the customer all along the life of the equipment at home. Some of the services will leverage access to information related to the product, its environment and its use; and some will provide specialized services (installation, maintenance, repair, etc.).

One example is boilers. The retailer has referenced different service providers according to their abilities and location. One SmartTag is attached on the boiler by the customer at home, by the manufacturer or by the retailer in-store and another SmartTag is attached inside the boiler by the manufacturer or by the retailer in-store. There are two more static SmartTags, one is attached on the boiler's burner and the second Smart-Tag is attached inside the filtering station upstream of the boiler, see Figure 9.7.

The environmental conditions at home are important to preserve the customer health and security, to control optimal conditions for the appliance operation (e.g. enough oxygen for proper combustion), and to detect malfunction of the appliance (e.g. carbon monoxide produced by the boiler) and optimize working conditions. Motivation for Home services use case is also that home equipment requires regular maintenance and many customers forget to proceed with this action; thus causing malfunction, shortening the life of the appliance, losing warranty, and losing insurance protection. From the retailer point-of-view warranty can only be set off in case of proper use

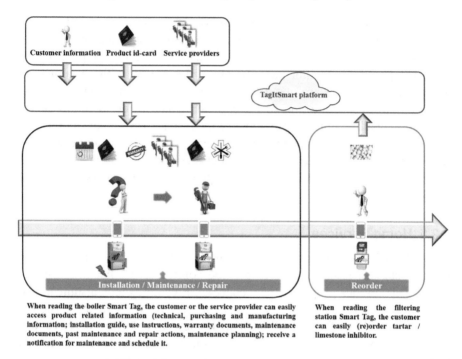

Figure 9.7   The idea of Home services use case.

and on time, maintenance of the equipment and the retailer is rarely able to verify this point. Quality and durability of the appliance are leveraged by good maintenance practices and enhance customer satisfaction. However, the retailer is rarely involved in this post-purchase process. Optimal use of equipment also enhances customer satisfaction.

9.3 Architecture

TagItSmart architecture was created based on the work done in previous projects like SocIoTal [6], available IoT system architecture reference models like IoT-A [7] and IoT platforms like FIWARE [8], Microsoft Azure [9] and EVRYTHNG [10]. Figure 9.8 presents a comprehensive view of TagItSmart architecture.

Based on this previous work, a list of identified required components for individual TagItSmart use cases were created. Then, they were combined into a common architecture, providing required functionality across all use cases. The following blocks were identified:

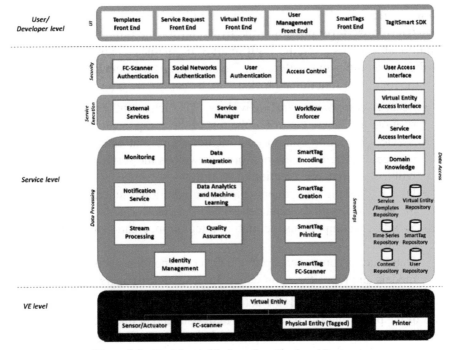

Figure 9.8 TagItSmart Platform functional architecture.

- **Service Management**: defines components needed to access, discover and execute services in the platform.
- **Virtual Entities**: defines components to work with virtual representations of the different objects defined in the use cases, from a CPG product to a sensor in a boiler.
- **SmartTags**: defines the components to manage the creation, scanning and management of SmartTags.
- **Security**: defines the components that will implement the authentication, authorisation, and any other security related aspects of the platform.
- **Domain Management**: defines components that are specific to a use case domain.
- **Utility Services**: defines utility components with services shared across use cases.
- **Application Development Tools**: defines the Software Development Kits (SDK) and tools needed to implement applications on top of the platform.

From the functional point of view, there will be a set of components at the **User/Developer level** providing the front end for the different components. Additionally, at this level, the TagItSmart SDK enables integration of specific TagItSmart features into different IoT platforms. At the **Service level,** we can find the following functional blocks:

- Security components dealing with aspects such as authentication, authorisation and access control to the rest of the components.
- Service Execution components include those that enable execution of services registered in the platform, as well as the service templates that will trigger dynamic creation of workflows.
- Data Processing components providing additional functionality to handle and work with the data generated in the platform.
- SmartTags components facilitating integration, creation and scanning of SmartTags.
- Data Access components provide the correspondent registries, semantic models and repositories on which TagItSmart will operate.

At the **Virtual Entity** level, the actual representations of the objects that are part of the platform provide access to their data and defined actions based on the semantic models. Each component is intended to define an integration strategy, mainly by the definition of an Open API. During the technical development in the correspondent work packages a reference implementation will be provided for the core innovative components (i.e. all the components directly related to the SmartTags and services), while others will be taken/integrated from already existing platforms.

9.4 Pilots and Trials

Validation of TagItSmart components is being performed in laboratory and controlled live environment, with the final validation planned to be done in live environment under real usage conditions. The goal of the trials is to validate requirements against individual components and services, while the focus of the pilot will be on the complete framework, the ease of use, end users' feedback, ease of use from the developers' point of view, integration with other systems, etc.

TagItSmart framework consists of a number of components required to enable the complete workflow of all envisioned use cases. However, not all individual components will be included in the validation during trials and pilots. Instead, the main focus of the trials will be on validation of the core components specific to TagItSmart, i.e., components involved in SmartTag scanning driven process, from the design and printing of the tags, to processing of data obtained from the tags together with the contextual information and finally delivering corresponding services.

The aim of the pilots will be on validation of functionality and performance of the system as a whole, not on the individual components. Further to the trial and pilots run by the project partners, additional validation will be done by the new partners joining the project through the open calls.

Operational trials for the different use cases have so far focused on different aspects. Some use cases have focused more on functional ink related issues (Brand protection), some on scanning of codes (Digital product), some on software component development (Life-cycle management), and some on preparations for pilot trials (Dynamic pricing and Home services). In the rest of this section we will present some sample results coming from the testing that has taken place already in the TagItSmart project. As the key enabler for the whole TagItSmart project are the SmartTags themselves, special effort has been given in the process of creating and scanning SmartTags in a way that sensing information can be reliably added to them and also reliably extracted from them.

9.4.1 SmartTag Creation

The main steps in the SmartTag creation process are the encoding of Smart-Tags to allow sensor information to be represented and the printing of the SmartTags themselves.

9.4.1.1 SmartTag Encoding

In TagItSmart project, Data Matrix [12] codes and QR codes have been used in the use cases. In the case of Data Matrix codes, in Digital product use case, different approaches how to integrate sensor information into existing 2D codes have been prototyped. Idea was to find the best combination of: 1) limitation of printing procedure and functional inks, 2) reading capability of mobile application, and 3) the cost of a solution. The best alternative so far has been to use a 2D code with a bar of functional ink overlaying part of the code, see an example in Figure 9.9. In addition, a square with functional ink was tested.

Encoding process itself is very easy when a rectangular symbol is used: data matrix is encoded based on the existing standards and sensor rectangle is added on top of the data matrix. Also, it is possible to encode a tag in the

Figure 9.9   2D code with sensor area.

reverse order, i.e. first, the sensor rectangle is encoded and then data matrix is layered on top of it. As these are most simple cases (and work in various printing systems), more complex encoding might be used for example in digital printing; in this case placing algorithm is used after the data matrix has been encoded. This algorithm replaces white cells at two last rows of the barcode with sensor ink. Information of the used sensor will be stored into (according to our TagItSmart architecture) the virtual entity counterpart of the product following the workflow shown in the figure below.

In the case of encoding QR codes, there are several issues associated with encoding tags that change with environmental conditions. These issues stem from the error correction mechanisms built into the QR code generation algorithms. The simplest approach to encoding a changing tag would be to simply design the two tags that are needed and then print the differences in disappearing or invisible inks that change under the same conditions. However, it is very difficult, and for some environmental conditions impossible, to change both inks under the same conditions. Therefore, the approach to encoding requires that only one type of ink (disappearing or invisible) be used. In the TagItSmart project, dual and mono coded tags have been developed. In dual coded tags, two tags are created. Each is designed in the normal way; one tag reads as the 'before' activation and the other reads as the 'after' activation. Before printing a carefully chosen set of dots on each tag are identified for printing with reactive inks. Both tags are printed side by side and the scanner will only read one of them. The mono tag approach is more complicated and restrictive. The design process must be allowed to choose some of the characters in the code.

It is also possible to encode SmartTags that are combinations of QR codes and images; this allows for embedding sensor information into SmartTags in cases a standalone use of a QR code would not allow for this (e.g. transitions between 2 different readable QR codes are not possible due to limitations in the way functional inks react).

9.4.1.2 Printing SmartTags

Different printing methods have been evaluated for their capability to print QR and Data Matrix codes with different functional inks on different paper based substrates. The concept for using functional inks in enhancing dynamic nature of 2D bar codes is that parts (e.g. some of the cells) of a 2D bar code are printed with a functional ink that reacts to the surrounding environment, e.g. temperature, humidity or light conditions. Changes in the environmental conditions cause the cells printed with a functional ink to disappear, to appear, or to change colour. This changes the physical layout of the 2D bar code thus also changing the information content (e.g. website address). Thereby the environmental conditions have an effect on the scanning process and resulting digital service. The services can also take into account the other user and context related data, such as user profile on the smartphone and GPS location of the smartphone. These context-aware features will further improve the value and quality of the services provided.

There is a range of colour-changing technologies, out of which the most popular and readily printable are thermochromic (changing colour with temperature) and photochromic (changing colour with sun or ultraviolet light, even camera flash light). Functional ink technologies are available at the moment mainly for analogue printing methods (flexography, screen printing, offset printing). For digital printing mostly, fluorescent inks are available. This is due to the large particle size of the colour-changing pigments, which present challenges in inkjet ink formulation. Functional inks can be used for printing 2D bar codes similar to regular printing inks, and detected by smartphones [11].

For printing functional inks both analog and digital printing methods are suitable. In analog printing (e.g. flexography, screen printing, offset) a physical master, a printing plate, is required for image reproduction thereby not making analog printing economically feasible for printing customized (individual) 2D bar codes. Analog printing is, however, capable of producing large batches in high speed with static printing layouts. Digital printing methods (e.g. inkjet, electrophotography), on the other hand, are suitable for printing customized, even personalized, 2D codes since the printing layout comes from a digital file without a physical master. There are inkjet printers available designed for printing variable data, such as Data Matrix and QR codes, for coding and marking applications. These printers are suitable for integration with other production and printing lines, and produce variable data on-the-fly. Manufacturers include Domino Printing (www.domino-printing.com) and Videojet (www.videojet.com). TagItSmart project partner Durst manufacturers label inkjet printers also suitable for printing variable data with 2D codes.

There are several different ink technologies available for analog printing methods, and those are readily usable with existing analog printing presses. For digital printing the availability is limited to only a few technologies, mainly UV or IR fluorescent inks.

There are, however, some limitations related also to functional inks for analog printing. The colours and colour changes available are limited. For example, with thermochromic inks the colour disappears when the temperature is higher than a specific limit. In some cases, it might be preferred to have an appearing ink instead of a disappearing one. Another issue is the difference between inks and indicators. For some cases, such as time-temperature monitoring (product at a particular temperature longer than a particular duration), there are only indicators available – not inks. Monitoring of time-dependent temperature changes is somewhat complicated and requires a logging capability that cannot be achieved with a single ink. This limits the usage scenarios since use of variable data is not possible due to the need to purchase ready indicators from the manufacturers.

To conclude, when selecting the functional performance of the SmartTag, it is important to understand the limitations and capabilities of the functional ink technologies and indicators, and select the functional performance based on their availability and properties. Also when variable data (e.g. individual 2D codes) are required for smart tag production, a proper selection of the printing methods is required due to the limited availability of the functional ink technologies for digital printing methods. In most cases hybrid solutions (combination of analog and digital printing methods) should work, where the 2D codes are printed with regular inks with digital printing, and the functional parts with analog printing or assembled from indicators.

Thermochromic inks

Thermochromic ink is a type of ink that changes colour with heat. This can make certain images appear (or disappear) as soon as the label or product

Figure 9.10 Data Matrix code with two functional inks. Code size is 3 cm x 3 cm. In the left picture both thermochromic inks can be seen (T < +31°C). In the middle picture black thermochromic ink has disappeared (+31°C < T < +47°C). In the right picture both black and red thermochromic inks have disappeared (T > +47°C).

goes above or below a certain temperature. These temperatures can vary from −10 to +65 °C. Temperature-sensitive inks come in two varieties: reversible and irreversible. With reversible thermochromic ink, the colour will revert when the temperature returns to its original level, whereas the colour remains constant after a change in temperature with irreversible thermochromic ink.

One example of a QR code that contains two flexography printed thermochromic inks is presented in Figure 9.10.

Photochromic inks
Photochromic materials change their colour when the intensity of incoming light changes. Most photochromics change from colourless to coloured upon exposure to UV light, and then fade back to colourless upon removal from the UV source. The normal wavelength of excitation is around 360 nanometres. Moreover, while sunlight works the best, a fluorescent black light, which emits near-UV light (320–400 nm), will usually work. Different dyes have different kinetics, meaning some will colour and fade quickly, while others will colour and then fade slowly. One example of an inkjet printed irreversible photochromic ink is presented in Figure 9.11.

NFC manufacturing
Printed Dopant Polysilicon Technology (PDPS) is used to produce the Printed Integrated Circuit (PIC) used in the NFC labels in TagItSmart project. The production has the following steps: PIC Production, wafer conversion (PIC bump and dice), dry inlay assembly and wet inlay conversion. The PICs are currently produced in a sheet-based process where 4 out of 8 layers are printed, see Figure 9.12.

Figure 9.11   Printed tag with an inkjet printed photochromic ink (red QR code). The red QR code appears when exposed to UV or sunlight.

Figure 9.12   The PICs are produced on stainless steel sheets.

After the conversion to wet inlays, the remaining steps are to place the labels on the final product. This is typically done by a converter or by the customer.

9.4.2 SmartTag Scanning

As part of the SmartTag scanning process, enablers that allow authenticating an FC-Scanner user as well as decoding the content in SmartTags have been developed.

9.4.2.1 FC-Scanner Authentication

Since SmartTags may carry information that can reveal personal aspects about an individual and/or objects they own, authenticating a user before they are allowed to scan an object with a SmartTag attached to it is of big importance. As an example, one can imagine the case where a person is wearing a t-shirt with a SmartTag that allows retrieving the temperature of that person. An authorised person (e.g. a doctor) should be allowed to scan such a SmartTag; an unauthorised person who just happened to pick up the doctor's FC-Scanner device should be blocked from doing so.

To account for scenarios like the one above, an authentication module for FC-Scanner devices has been developed that based on mobility data of the owner, is able to "on the fly" authenticate a user automatically without requiring any intervention or direct input from the rightful owner, as the module learns to identify this rightful owner.

The TagItSmart platform can leverage the outcome of the authentication module to take remedial actions, such as "lock" the FC-Scanner device, ignore SmartTag scanning events coming from it and also notify the rightful user (through other devices registered under his ownership) about this. The architecture of the authentication system for TagItSmart is presented in Figure 9.13.

9.4.2.2 Decoding SmartTags

For all (Data Matrix, QR code, QR code together with image) types of SmartTags currently successfully encoded, the decoding mechanisms have also been developed. In the case of a SmartTag consisting of a QR code together with an image, a special TagItSmart decoding application had to be developed combining an off the shelf QR code decoder together with a custom made image recognition module. The decoding performance has been very promising with successful decoding of a SmartTag being able even for very small SmartTags (down to 2cm × 2cm in size and even lower), with high speed (lower than 2.5 seconds) and under low lighting conditions, as low as 35 lux. Figure 9.14 presents the SmartTags being successfully decoded when attached to real-life products.

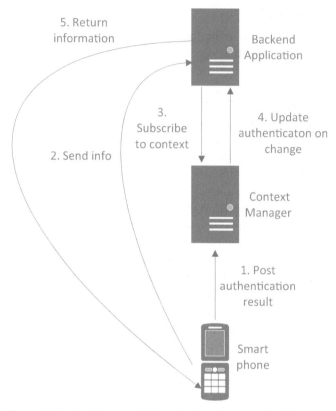

Figure 9.13　Architecture of the authentication system for TagItSmart.

In a similar manner, NFC based SmartTags produced by the project are capable of allowing for scanning reliability of higher than 99% with response time lower than 0.5 seconds, as long as the scanning distance matches the requirements of near field communications (lower than 2cm).

9.4.3 Service Offerings Leveraging the TagItSmart Platform

The various enablers developed in the TagItSmart project allow for many services to be offered leveraging the information that is extracted from SmartTags. This is still an ongoing process given the early phase of the project and the need to develop first the enablers that allow getting in and out of SmartTags the information that more advanced service offerings rely

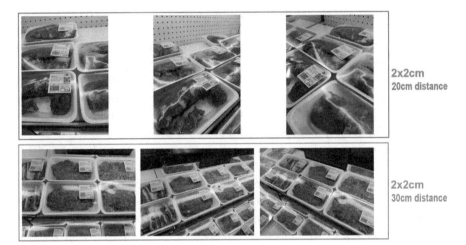

Figure 9.14 Examples of SmartTags being successfully decoded when attached to real-life products.

upon. In the context of the Dynamic pricing use case, the service scenario of presenting different prices for the same product based on how close this is to its expiry date (as this is reflected by the attached SmartTag) has already been tested successfully, illustrating the successful interworking and integration of encoding, printing, decoding and decision making for this scenario (see Figure 9.15).

Figure 9.15 Dynamic pricing scenario; different price calculated and displayed based on SmartTag information.

9.4.4 User Experience on Use Cases

Use cases, developed in TagItSmart project, have been evaluated with Finnish consumers by using Owela platform (Open Web Lab, http://owela.fi). Owela has been developed at VTT in 2007 for user centric qualitative studies and it supports active user involvement in the innovation process from the early ideas to piloting and actual use [12]. Owela can be utilized to involve users in all phases of a development process for an innovation.

Discussion in Owela was open during two weeks from January 2017. Totally, 45 Finnish consumers participated the discussion and gave 341 comments to the discussion. Fifty-three percent of the participants were female, and 47% of male. Age distribution varied between 17–79 years, the average age of the participants was 49 years. The use cases were presented to the participants as stories to understand the idea of the use case. After reading the stories, participants discussed about their interest towards the different concepts. Based on the results, it was clear that people in Finland are familiar with 2D codes. All the participants had seen them e.g. in the packages, advertisements and tickets. Three out of four had read 2D codes with their smartphones but only 17% said that they read the codes regularly.

Consumers found all of these concepts interesting. Seventy-nine percent of the participants evaluated it to be interesting to read the codes from local producers' products and to create an interactive relationship with them. To utilize the codes in fast moving consumer goods, like meat packages, was in the interest of 55% of the participants. Finnish consumers did not see so much potential in using the codes in cheap products and they were not concerned about the origin and safety of the products. That was also the case in the authenticity of the products, but still consumers saw a lot of potential in this case. Eighty-five percent of the participants evaluated the authenticity of the products interesting.

In addition, we got lots of information about the advantages of the TagItSmart concepts, and also about consumers' concerns related to them. First, even if consumers are used to seeing and using 2D codes, they have often been disappointed with the content they can receive by reading the codes. They expect something more – high quality content that offers them additional value. In addition, consumers were also interested in the possibility of interactivity. There were most potential in using codes in unique and/or personalized products, ensuring food safety of the authenticity of the products and also in creating interactive connection with producers.

Consumers were suspicious about if it is too easy to counterfeit the codes, recycling of the tags and also privacy issues related to the novel IoT services.

It is essential that the consumers can trust the service and service provider. It is important to take into account when TagItSmart solutions will be brought to the market, how the services are communicated so that consumers understand the difference between traditional codes and functional ones, and also to realise the added value and ensure the trust.

9.5 Conclusion

The activities performed in TagItSmart project so far, prove that there is a great business potential for connected FMCG products, across a number of domains. All industrial partners in the consortium have identified use cases of interest and are now proceeding to validate them in the pilots. Further to this, based on interaction with various stakeholders as well as External Industry Advisory Board, the proposed use cases were validated as being of interest to a wider number of organizations. On the consumer side, even if they are used to seeing and using 2D codes, they have often been disappointed with the content they can receive by reading the codes. They expect something more – high quality content that offers them additional value. In addition, consumers were also interested in the possibility of interactivity and these are the features TagItSmart can and is planning to offer.

Technology is available to produce novel services for consumers and stakeholders by utilising functional inks and combining them with digital solutions. A number of challenges related to printing codes using functional ink and even logistics of printing on-site or transporting codes sensitive to temperature or other environmental parameters to the labelling facility, have to be addressed before large scale uptake becomes possible.

The process of coming up with solutions that will enter the market is iterative, and potential users are involved in different phases. It is important to take into account how the services are communicated so that consumers understand the difference between traditional codes and functional ones, and also to realise the added value.

Acknowledgements

This work was supported by EU; the TagItSmart project (Grant Agreement No 688061) is part of Horizon 2020 program. In the project, we have altogether 15 partners from Serbia, Finland, United Kingdom, France, Sweden, Romania, Netherlands, Italia, Austria and Spain. We want to thank all the partners for their co-operation during the project.

References

[1] Consumer Goods Forum, 'Rethinking the Value Chain: new realities in collaborative business', 14 December 2015. [Online]. Available: https://www.uk.capgemini.com/resources/rethinking-the-value-chain-new-realities-in-collaborative-business.

[2] P. Conroy, K. Porter, R. Nanda, B. Renner, A. Narula, 'Consumer product trends: Navigating 2020', 25 June 2015. [Online]. Available: http://dupress.com/articles/consumer-product-trends-2020/.

[3] C. Links, 'Consumers want Smart Systems Not Smart Things', Green-Peak, 8 January 2016. [Online]. Available: http://www.sensorsmag.com/networking-communications/consumers-want-smart-systems-not-smart-things-20558.

[4] Consumer Goods Forum, 'Rethinking the Value Chain: new realities in collaborative business', 14 December 2015. [Online]. Available: https://www.uk.capgemini.com/resources/rethinking-the-value-chain-new-realities-in-collaborative-business.

[5] SeeDiscover, 'Behind The Numbers: Growth In The Internet Of Things', 23 October 2015. [Online]. Available: http://www.seediscover.com/behind-the-numbers-growth-in-the-internet-of-things/.

[6] SocIoTal, 'Creating a socially aware and citizen-centric Internet of Things!', [Online]. Available: http://sociotal.eu/.

[7] IoT-A, 'Internet of Things – Architecture IoT-A Deliverable D1.5 – Final architectural reference model for IoT v3.0', [Online]. Available: https://www.researchgate.net/publication/272814818_Internet_of_Things_--_Architecture_IoT-A_Deliverable_D15_-_Final_architectural_reference_model_for_the_IoT_v30.

[8] Fiware. [Online]. Available: https://www.fiware.org/.

[9] Microsoft Azure. [Online]. Available: https://azure.microsoft.com/en-us/?v=17.14.

[10] EVRYTHNG. [Online]. Available: https://www.evrythng.com/.

[11] L. Hakola, H. Linna, 'Detection of printed codes with a camera phone', Conference proceedings. 32nd International Research Conference Iarigai, 2005, Porvoo.

[12] P. Friedrich, P., 'Web-based co-design: Social media tools to enhance user-centred design and innovation processes', Espoo 2013. VTT Science 34. 185 + 108 p.

Index